# BASIC GROWTH ANALYSIS

## Plant growth analysis for beginners

---

**RODERICK HUNT**

BSc, PhD, CBiol, MIBiol

*Independent Research Worker, Natural Environment Research Council,*
*Unit of Comparative Plant Ecology*
*Honorary Lecturer in Botany, University of Sheffield*

LONDON
UNWIN HYMAN
Boston Sydney Wellington

Published by the Academic Division of
**Unwin Hyman Ltd**
15/17 Broadwick Street, London W1V 1FP, UK

Unwin Hyman Inc.,
8 Winchester Place, Winchester, Mass. 01890, USA

Allen & Unwin (Australia) Ltd,
8 Napier Street, North Sydney, NSW 2060, Australia

Allen & Unwin (New Zealand) Ltd in association with the Port Nicholson Press Ltd,
Compusales Building, 75 Ghuznee Street, Wellington 1,
New Zealand

First published in 1990

---

**British Library Cataloguing in Publication Data**

Hunt, Roderick *1945–*
Basic growth analysis
1. Plants. Growth. Analysis. Quantitative methods.
I. Title.
581.3′1

ISBN 0-04-445372-8
ISBN 0-04-445373-6 pbk.

---

**Library of Congress Cataloging in Publication Data**

Hunt, Roderick, 1945–
Basic growth analysis : plant growth analysis for beginners/ Roderick Hunt.
p. cm.
Bibliography: p.
Includes index.
ISBN 0-04-445372-8 (alk. paper).—ISBN 0-04-445373-6 (pbk. : alk. paper)
1. Growth (Plants)—Experiments. 2. Growth (Plants)—Mathematical models. 3. Growth (Plants)—Measurement. I. Title.
QK731.H85 1989
581.3′1′015118—dc20 89-16462
CIP

---

Typeset in 10 on 12 point Palatino by MCS, Salisbury, England and printed in Great Britain by Cambridge University Press

TITLES OF RELATED INTEREST

*Class experiments in plant physiology*
H. Meidner (ed.)

*Comparative plant ecology*
J. P. Grime, J. G. Hodgson & R. Hunt

*Crop genetic resources*
J. T. Williams & J. H. W. Holden (eds)

*Introduction to vegetation analysis*
D. R. Causton

*Introduction to world vegetation (2nd edition)*
A. S. Collinson

*Light & plant growth*
J. W. Hart

*Lipids in plants and microbes*
J. R. Harwood & N. J. Russell

*Plant breeding systems*
A. J. Richards

*Physiology & biochemistry of plant cell walls*
C. Brett & K. Waldron

*Plant development*
R. Lyndon

*Plants for arid lands*
G. E. Wickens, J. R. Goodin & D. V. Field (eds)

# BASIC GROWTH ANALYSIS

# *Preface*

This handbook is intended as an introductory guide to students at all levels on the principles and practice of plant growth analysis. Many have found this quantitative approach to be useful in the description and interpretation of the performance of whole plant systems grown under natural, semi-natural or controlled conditions.

Most of the methods described require only simple experimental data and facilities. For the classical approach, GCSE biology and mathematics (or their equivalents) are the only theoretical backgrounds required. For the functional approach, a little calculus and statistical theory is needed. All of the topics regarding the quantitative basis of productivity recently introduced to the Biology A-level syllabus by the Joint Matriculation Board are covered.

The booklet replaces my elementary *Plant Growth Analysis* (1978, London: Edward Arnold) which is now out of print. The presentation is very basic indeed; the opening pages give only essential outlines of the main issues. They are followed by brief, standardized accounts of each growth-analytical concept taken in turn. The illustrations deal more with the properties of well-grown material than with the effects of specific environmental changes, even though that is where much of the subject's interest lies.

However, detailed references to the relevant parts of more comprehensive works appear throughout, and a later section on 'Interrelations' adds perspective. Some 'Questions and answers' may also help to show what topics will arise if the subject is pursued further.

RH
*Sheffield, 1989*

# Acknowledgements

The permissions of the authors and publishers of the various illustrative examples are gratefully acknowledged. Though not specifically connected with this project, Drs D. R. Causton and G. C. Evans stand foremost among the many friends who have been of personal assistance to these endeavours over the years. Mrs A. M. N. Ruttle has provided incomparable secretarial support and Professor J. N. R. Jeffers kindly commented on the manuscript.

The Unit of Comparative Plant Ecology is part of the Terrestrial and Freshwater Sciences Directorate of the Natural Environment Research Council.

We are grateful to the following individuals and organizations who have kindly given permission for the reproduction of copyright material (figure numbers in parentheses):

R. Sattler and Springer-Verlag (Table 1.3); British Ecological Society (Table 2.1); D. R. Causton and Edward Arnold (2.2, 4.3, 4.5, 4.11); W. J. Leverich and the University of Chicago Press (Tables 2.2, 3.2); Annals of Botany Company (2.3, 2.4, 4.4, 4.6, 5.2, 5.4, 8.5, Tables 4.7, 8.3); The Editor, *The New Phytologist* (3.1, 3.2, 3.3, 4.1, 5.1, 5.9, 8.1, 8.2); T. Ingestad and The Editor, *Physiologia Plantarum* (Tables 3.1, 5.3); E. Ashby and the Annals of Botany Company (3.4); S. Arvidson and Ellis Horwood Ltd (3.5); The Editor, *Journal of General Microbiology* (3.6); The Editor, *Journal of Ecology* (Tables 4.1, 8.1); L. T. Evans and Cambridge University Press (Tables 4.4, 4.6, Figure 6.1); C. B. Johnson and Butterworths (Table 4.5); Cambridge University Press (4.7); J. K. A. Bleasdale and Macmillan Publishers Ltd (4.8); The Editor, *Journal of the British Grassland Society* (4.9); F. H. Whitehead and The Editor, *The New Phytologist* (4.10); J. Coombs, © 1982 Pergamon Press PLC (Table 5.1); The Editor, *Australian Journal of Agricultural Research* (Table 5.2); R. F. Williams and the Annals of Botany Company (5.3); R. F. Williams and Cambridge University Press (Table 5.4); P. J. Welbank and the Annals of Botany Company (5.6); D. P. Stribley and The Editor, *The New Phytologist* (5.7); P. J. Dudney and the Annals of Botany Company (5.8); J. Květ (Table 6.1); J. Květ and The Editor, *Photosynthetica* (6.2, Table 6.2); D. R. Tottman (7.1); The Editor, *Physiologia Plantarum* (8.3); R. M. Shibles and Cambridge University Press (8.4); Institute of Terrestrial Ecology (Ch.9 Checklist).

# *Contents*

# CHAPTER ONE
# *Introduction*

# GROWTH

All living organisms are, at various stages in their life history, capable of growth. Given suitable conditions, this can mean change in size (Fig. 1.1), change in form and/or change in number. These three processes together form an important part of the phenomenon of life. Among natural systems they help to distinguish the living from the non-living though, in a sense, many non-living systems also grow. Crystals, river deltas and volcanic cones can change recognizably within human time-scales.

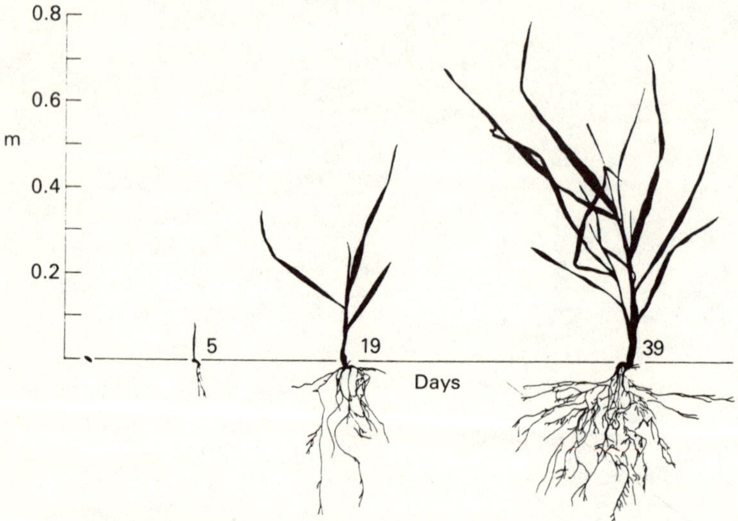


**Figure 1.1 Silhouettes of barley** Great increases in size may occur during the first few weeks of growth in a productive environment. (Photograph: G. Woods.)

But even within self-reproducing biological organisms, a precise definition of what is meant by 'growth' is not easy. Definitions range from unequivocal statements about change in specified dimensions to the abstract state of affairs in which the verb 'to grow' means nothing more than to live or even to exist. *The Concise Oxford Dictionary* gives it as (1) 'develop or exist as living plant' and (2) 'increase in ... size, height, quantity, degree, power, etc.'. For plant growth analysis, let us lean towards the second definition. We can then say that growth describes irreversible changes with time which are mainly in size (however this may be measured), often in form, and occasionally in number.

A plethora of separate scientific publications have appeared which quite specifically involve some aspect of plant growth. To keep our story short we must ignore whole fields of study, from cell division, through growth regulation and morphogenesis, to environmental physiology and agronomy. None of these topics will receive attention here in their own right. The scope of this work is simply to introduce the quantitative analysis of a series of observations on the growth of plant organs, whole individuals, populations and communities.

What form may such observations take? Let us see some hard data straight away. A series of classical experiments was performed at Poppelsdorf, West Germany, in the 1870s by Kreusler and co-workers (reference in Table 1.1). The results demonstrated that the growth of an annual plant under natural conditions followed a course that has since been recognized as typical of many. Table 1.1 contains a selection of Kreusler's data, showing the pattern of increase with time in mean dry weight and leaf area per plant in *Zea mays* (maize) cv. 'Badischer Früh' grown in 1878.

This set of observations was the culmination of several years' work with different varieties of maize. Each measurement is a mean taken from many individual observations. I have converted the units of total leaf area from $cm^2$ to $m^2$, otherwise the values are reproduced exactly in the form given by Kreusler.

These primary data are of high quality and they still remain relevant to the process of describing and interpreting the growth of whole plants. Remarkably, they were collected long before modern methods of experimental design and sampling had evolved. What can be seen on first inspection of them?

This particular set forms a series of 17 observations made, with one exception, at weekly intervals throughout a whole growing season. We see that both measures of plant size span a substantial range. We see also that it was not feasible to determine total leaf area per plant until the fourth sampling occasion. However, there we must leave this section, because any more than these preliminary observations would come into the category of plant growth *analysis*.

**Table 1.1 Kreusler's maize data** Five varieties of maize were grown at Poppelsdorf in 1875. Two of these, including variety 'Badischer Früh', were also grown in 1876; 'Badischer Früh' was again grown alone in 1877 and in 1878. The data obtained for 1878 are given here. Each mean value of dry weight is based upon 40 to 120 samples, and of leaf area upon 20 to 120 samples. (From: U. Kreusler *et al.* (1879). *Landw.Jbr* 8, 617.)

| Date of harvesting | Day in the year | Mean total dry weight per plant (g) | Mean total leaf area per plant ($m^2$) |
|---|---|---|---|
| 20 May 1878 | 140 | 0.3282 | n.a. |
| 28 May | 148 | 0.328 | n.a. |
| 4 June | 155 | 0.287 | n.a. |
| 11 June | 162 | 0.255 | 0.00179 |
| 18 June | 169 | 0.308 | 0.00292 |
| 25 June | 176 | 0.637 | 0.01244 |
| 2 July | 183 | 2.319 | 0.04192 |
| 9 July | 190 | 4.654 | 0.07622 |
| 16 July | 197 | 9.019 | 0.1301 |
| 23 July | 204 | 20.001 | 0.2136 |
| 30 July | 211 | 34.557 | 0.2805 |
| 6 August | 218 | 57.587 | 0.3384 |
| 13 August | 225 | 70.095 | 0.3047 |
| 20 August | 232 | 85.165 | 0.3025 |
| 27 August | 239 | 111.649 | 0.2976 |
| 3 September | 246 | 124.760 | 0.2684 |
| 10 September | 253 | 121.990 | 0.2387 |

n.a. = not available

Obvious though it may seem, the first stage in the analysis of data such as those given in Table 1.1 is to point them out.

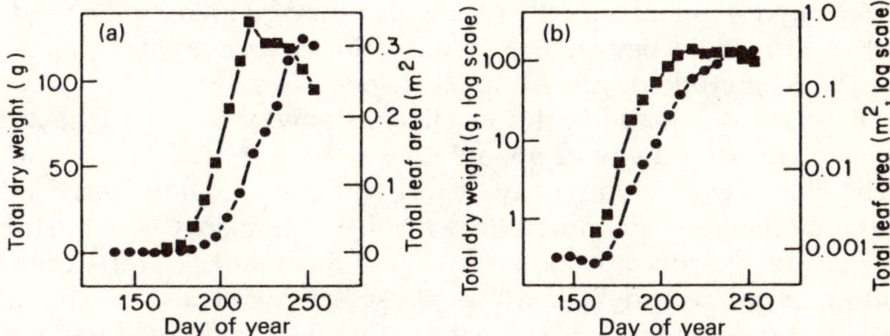


**Figure 1.2 Arithmetic and semi-logarithmic plots** Data from Table 1.1 are plotted here on (a) an arithmetic scale and (b) a semi-logarithmic scale to illustrate the value of the logarithmic transformation.

When this is done (in Fig. 1.2a) we encounter a first difficulty. Because the changes in dry weight over the whole period are of the order of 370-fold, very little of the first phases of development is revealed in this simple plot of dry weight against time on an arithmetic scale. If the data are transformed to (natural) logarithms, we can see more clearly what is happening (Fig. 1.2b). There is no special reason as yet why natural, rather than common, logarithms should be used for this purpose (or indeed some other transformation such as square root), but the natural logarithms will turn out to be essential later.

With the transformation, we see that the plant shows no change in dry weight for the first ten days or so. Then it actually loses weight until about 20 days have passed. Here is an example of another difficulty encountered in defining growth: no increase in weight has occurred but there has been considerable differentiation of leaf tissue in the young seedling (at the expense of total dry weight). From about day 170 in the year, the newly differentiated leaves begin to contribute substantially to carbon assimilation. A so-called grand period of growth begins in which the unfolding of new leaves and the increase in total dry weight occur continuously. The plants flower at about day 205. Thereon, an increasing proportion of assimilate is directed into the developing ear or cob with a corresponding tendency for the lower leaves to atrophy. Finally, at about day 246, net dry weight increase ceases although growth in the sense of a continuing partition of dry weight into ears continues.

This pattern of growth is quite general to annual plants grown in a productive environment, though there is great variation in the magnitude of the dry weight values, in the symmetry of the curve and in the time scale which it occupies. In perennial plants the pattern is similar at first but later, at least in a temperate climate, dry weight increase proceeds in a series of annual steps which may be linked by periods of negative growth in between. Naturally, the environmental conditions affect the magnitude of growth at all stages.

Preliminary synopses like this can take us only so far. Over the last 70 years an additional body of quantitative techniques has been build up which allows experimenters to derive important comparative information about the undisturbed growth of whole plants under natural, semi-natural or artificial conditions. These techniques require only the simplest of primary data, such as those described above, and have collectively become known by the informal title 'plant growth analysis'.

The origin of this activity has been chronicled by EVANS (1972, pp. 189–205). The References section cites this book, and other major sources, in full. The techniques of plant growth analysis are, above all, powerful comparative tools. They have been developed to negate, as far as possible, the inherent differences in scale between contrasting organisms so that their performances may be compared on an equitable basis.

Table 1.2 lists the rates of dry weight increase (irreversible growth in size) for a variety of organisms grown under favourable conditions. This shows the utility for comparisons of one of the chief concepts of plant growth analysis. The rates of growth range below 10% per day in large trees, through intermediate rates in herbaceous plants, algae, fungi and microorganisms, to rates exceeding 20 000% per day in an anti-*Escherichia coli* phage. Despite much variation within groups, one broad conclusion is clear: the larger and more complex the organism, the lower the rate of dry weight increase that is possible when expressed on a percentage basis. This trend is thought to be due to the increased morphological and anatomical differentiation which is necessary to sustain the life of large systems. This differentiation leads to translocatory pathways of increased length between the point of entry of raw materials into the organism and the site of its nucleoprotein replication. However, the point that the table really illustrates is that although the

**Table 1.2 Rates of dry weight increase** Percentage dry weight increases in organisms grown under favourable conditions. (From: HUNT, p. 2).

| Group | Organism | % $day^{-1}$ |
|---|---|---|
| Bacteriophage | Anti-*Escherichia coli* phage | 20400 |
| Bacterium | *Escherichia coli* | 4750 |
| Virus | Tobacco mosaic virus | 2210 |
| Yeast | *Willia anomala* | 1400 |
| Fungus | *Aspergillus nidulans* | 860 |
| Algae | *Chlorella* (T × 7115) | 620 |
| | *Anabaena cylindrica* | 74 |
| Angiosperms – herbaceous | *Poa annua* | 38.6 |
| | *Helianthus annuus* | 29.0 |
| | *Nardus stricta* | 10.1 |
| Angiosperms – woody seedlings | *Fraxinus excelsior* | 12.8 |
| | *Acer pseudoplatanus* | 4.85 |
| Gymnosperms – seedlings | *Picea abies* | 6.00 |
| | *Picea sitchensis* | 3.14 |

differences in organization between these groups could scarcely be greater, calculations made in this particular way allow fair quantitative comparisons to be drawn.

Of course, other fields of study also aim to uncover quantitative detail about the growth of plants. Studies of plant photosynthetic production – the applied aspects of photosynthesis research – investigate the plant/environment relationship at the level of the leaf, leaf segment and chloroplast. In comparison, plant growth analysis suffers the disadvantage of providing no deeply mechanistic information about the environmental responses of plants, even though valuable clues may sometimes emerge. On the other hand, the great advantage of many of the quantities involved in plant growth analysis is that they provide accurate measurements of the sum performance of the plant integrated throughout the whole undisturbed plant and across substantial intervals of time. To predict this from the starting point of purely physiological observations can involve many dangerous assumptions. In a sense, plant growth analysis judges the system more by results than by promises.

## 'CLASSICAL' AND 'FUNCTIONAL' APPROACHES

A division between two distinct approaches to plant growth analysis evolved in the 1960s. The above names were first used by D. R. Causton, but others have used the term 'dynamic' for what we shall call the functional approach. However, the terminology is relatively unimportant provided we realize that one approach necessarily involves the use of fitted curves and the other does not. The word 'functional' thus appears in the mathematical sense of a relation between variables rather than in the physiological sense of a mode of action or activity.

In the classical approach, the course of events is followed through a series of relatively infrequent, large harvests (with much replication of measurements). The main summaries of the approach appear in the books by EVANS and CAUSTON & VENUS.

In the functional approach, harvests supply data for curve-fitting; they are smaller (there is less replication of measurements) but more frequent. The main summaries of the approach appear in the books by CAUSTON & VENUS and HUNT.

The two approaches are not mutually exclusive if time and space are no object (harvests may be large *and* frequent), but it is not often that such a scheme makes the most efficient use of the material available. Hence, in most cases, the experimenter must decide in advance which approach to take since this will influence the design and execution of the experiment. Fortunately, the two approaches share many practical prerequisites. EVANS (pp. 6–185) gives a full survey of them.

Basically, we assume that experimenters grow plants in order to test an hypothesis. Possibly we wish to show that one particular environment or management practice is or is not more suitable for a particular plant than another; or we may wish to compare the performances of different species or varieties grown under the same conditions; or we may just wish to explore and quantify the growth of a new experimental subject. In any of these cases, a programme of activity can probably be designed that will give us at least some of the information that we are looking for through the medium of one or more of the approaches to plant growth analysis.

Although the design of experiments proper is outside the scope of this book, the problems involved can be set down simply: Can we be sure of selecting the right experimental material? Can we use it in such a way that any differences in growth genuinely reflect the treatments applied? Can we sample the material in such a way as to get down on to paper an accurate version of these differences? Can we handle these data in a way that illuminates our understanding of the real events that have occurred and provides the basis for the acceptance or rejection of our hypotheses (Table 1.3)?

**Table 1.3 The hypothetico-deductive method of science** Hypotheses cannot be proved, but they can be disproved. The scheme shown here is a considerable simplification of what happens in practice at all levels, especially the upper ones. The whole process also recycles frequently. (From: R. Sattler (1986). *Biophilosophy*. Berlin: Springer-Verlag.)

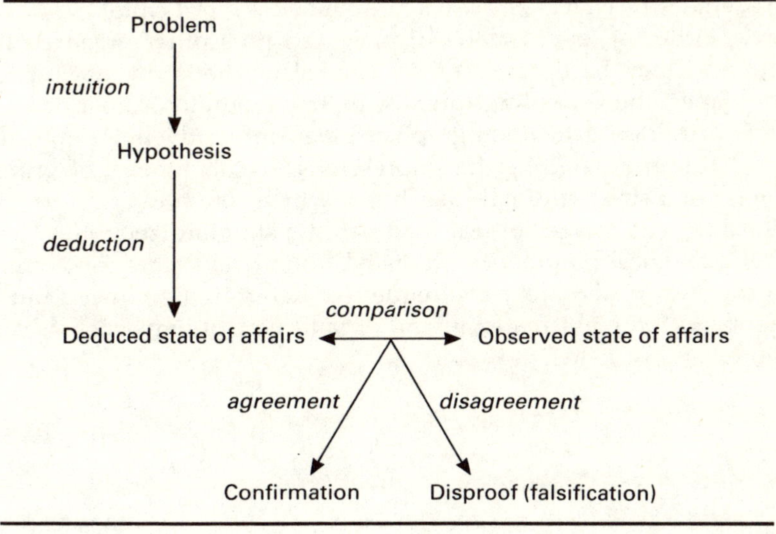

The stock-in-trade of plant growth analysis is a collection of simple primary data, the measured quantities upon which subsequent analyses depend. These may be determined either for the whole plant or for different sections such as roots, stems and leaves, as required.

For **fresh weight**, it is important to maintain standard moisture conditions before and during measurement. For **dry weight**, under- and over-drying must be avoided. A reliable system for measuring **length** is not difficult to devise but **volume** must be determined either by calculation, if the geometry of the plant is simple, or, if not, by the displacement of water (employing Archimedes's principle).

**Area**, either of leaves alone or of leaves plus other green parts, is useful and may be determined by several methods: by tracing on to graph paper, then counting squares; by tracing on to ordinary paper or contact-printing on to photographic, blueprint or dyeline paper, then cutting out and weighing the impressions; by planimetry of prints or tracings; or using automatic machines which use photocells or video cameras to scan images of leaf material in a standardized way.

Methodological handbooks are listed under the References, together with reviews of laboratory techniques for determining mineral nutrient contents within plant material, and organic constituents such as carbohydrates and proteins.

By international agreement all scientific work is now conducted using the SI system of units (*Système International d'Unités*). The standard unit of length is the metre and of mass the kilogramme. Multiples or fractions of units are restricted to steps of one thousand.

For plant growth analysis, this means that we use millimetres, metres and kilometres for length (and hence area) and microgrammes, milligrammes, grammes and kilogrammes for mass. For length and area, the very useful centimetre, square centimetre and square decimetre must be sacrificed to a good cause. Some workers feel that even this loss is slight compared to the difficulties which arise when considering volume. Here, no unit intervenes between the cubic millimetre and the cubic metre, a thousand million times larger. The SI units of energy (joule), power (watt) and customary temperature (degree Celsius) can be adopted with few, if any problems.

The SI unit of time is the second, but this is of much more use to physical science than to biology. In plant growth analysis, as in some other fields, the processes studied operate on a longer time scale than this and everyday units such as days and weeks have been retained, the only exception being in short-term work where processes are under more or less continuous observation. Table 1.4 lists the dimensions and SI units of the quantities most commonly needed in plant growth analysis.

**Table 1.4 Système International (SI) units** (for plant growth analysis).

| Quantity | Dimensions | SI units |
|---|---|---|
| Number | N | number |
| Mass | M | kg, g, mg, $\mu$g |
| Length | L | km, m, mm |
| Time | T | s |
| Area | $L^2$ (=A) | $m^2$, $mm^2$ |
| Volume | $L^3$ | $m^3$, $mm^3$ |
| Density | $M\ L^{-3}$ | kg $m^{-3}$ |
| Temperature | $\theta$ | °C or °K |
| Radiant energy | H (or $M\ L^2\ T^{-2}$) | J |
| Radiant flux | $H\ T^{-1}$ | W |
| Radiant flux density | $H\ L^{-2}\ T^{-1}$ | W $m^{-2}$ |

# NOTATION

In any field of scientific study a consistent notation is a great advantage, especially where equations and mathematical expressions abound. Notation in plant growth analysis had a fragmentary evolution until EVANS attempted the first comprehensive and coherent system of notation.

In principle, Evans's system is followed in this book. However, some of its admirable rigour has been eroded in the interests of preserving at least some links with other conventions, a policy also followed by CAUSTON & VENUS and by HUNT. The principles followed are these:

(a) Contractions of names: roman capitals (or small capitals); e.g. RGR, relative growth rate
(b) Measured quantities: italic capitals; e.g. *W*, total dry weight
(c) Derived quantities: bold capitals; e.g. **R**, relative growth rate
(d) Parameters of equations: italic lower case; e.g. constants *a*, *b*
(e) Distinguishing subscripts: roman, suffix position; e.g. $R_{\mathrm{W}}$, root dry weight
(f) Subscripts defining time: suffix position (prefix in EVANS); e.g. $t_1$, $W_1$, initial time and total dry weight
(g) Conventional signs and mathematical symbols: roman or italic, upper or lower case as required; e.g. W, watt; $\log_e$, natural logarithm; *P*, probability

A special synopsis table at the end of the book lists all contractions, symbols, expressions, formulae and units used. Note that the measured quantity, time, receives the conventional symbol *t* and not the more logical T. This usage can be defended if we distinguish between **variates** (statistical quantities which have a particular value for each member of a supposedly homogeneous population, with a particular frequency distribution of these values) and **variables** (quantities also able to assume different values, but capable of accurate representation on any one occasion). Most measures of plant size are variates, but time is a variable. However, some quantities, e.g. temperature, can be either.

One object of this book is to point out the opportunities that the increasing availability and accessibility of computing power is provided for those who wish to study plant growth. Though some parts of the functional approach can be tackled with nothing more than a set of logarithmic tables, I assume that readers will have some form of computing support available, which will also assist in many aspects of the classical approach. This is why most of the formulae are laid out as though for a computer program, rather than being displayed in the more usual style.

The computing support might consist of a programmable pocket calculator, a personal computer, a multimillion-pound institutional computer, or anything in between. It is obviously impossible to issue any specific guidance on computational details, particularly in a field which is changing so rapidly. Instead, my strategy is to outline the biological, mathematical and statistical characteristics of the various concepts within plant growth analysis, leaving the reader to implement them in his/her own particular way. The benefits to be gained are such that it is always worth while making a careful assessment of local computing possibilities. Even if modest, they can always be used to the full and in the most advantageous manner.

Good introductions to the important processes of hypothesis-generating, variability of material, sampling of populations, replication, randomization and measurement are listed in the References section, where there is also mention of introductory books on mathematics and on many of the necessary statistical techniques, including tests of significance, analysis of variance, correlation and regression.

It goes without saying that the analytical and statistical techniques to be applied should be firmly in the experimenter's mind before practical work is begun. Looking round for some way of 'doing the stats' after data capture is complete is a risky process since the value of a lengthy experiment can dissolve in seconds with the realization that the hard-won data are not in a form amenable to analysis. Plan the *whole* experiment first. If this is done not only is more efficient use made of time and manpower but the job is actually easier since a smooth flow of data can be arranged all the way from the bench to the write-up.

Now it is time to introduce the central concepts of plant growth analysis. One major subdivision may immediately be made. This is according to level of organization.

Readers will be familiar with the modern subdivision of biology, not by taxonomic or even by functional criteria, but by organizational structure. This approach implies a graded series which ascends in complexity from the molecule to the biosphere (Table 1.5).

**Table 1.5 Levels of organization** Appropriate concepts are indicated.

| Level of organization | Growth-analytical concepts |
|---|---|
| Biosphere<br>Ecosystem<br>Community<br>Population | Growth analysis of plant stands |
| Organism<br>Organ<br>Tissue<br>Cell<br>Organelle<br>Molecule | Growth analysis of individuals and components |

Concepts in plant growth analysis fit this scheme well since a major historical and practical threshold occurs at the interface of the organism and the population. However, the organism (and below) and the population (and above) form two halves of a scheme which remains recognizably similar throughout. There is thus an integrated scheme of activity in this area which is accessible to uniform coverage by an introductory handbook.

So, the division between studies concerning plants grown as whole, spaced individuals and plants grown as natural, semi-natural or agricultural populations or communities forms an important distinction within plant growth analysis. Following this introduction there is a standardized short treatment of important points regarding each of the quantities involved.

Fundamental to the whole field is a series of sequential measurements of plant size, form of number – the primary data. From these, one or more of five principal types of derived quantity can be constructed:

I *Absolute growth rates*. These are simple rates of change involving only one plant variate and time, examples being the whole plant's rate of dry weight increase, or the rate of increase in number of roots per plant.

II *Relative growth rates*. These are more complex rates of change, but still involve only one plant variate and time, an example being the whole plant's rate of dry weight increase per unit of dry weight.

III *Simple ratios*. These involve the ratio between two quantities; they may either be ratios between two like quantities, such as total leaf dry weight and whole-plant dry weight, or ratios between two unlike quantities, such as total leaf area and whole-plant dry weight. Sometimes, one of the quantities involved in the ratio is not a plant variate at all, such as the ground area per sample of crop.

IV *Compounded growth rates*. These are rates of change involving more than one plant variate, such as the whole plant's rate of dry weight increase per unit of its leaf area. Again, one of the variates may not be a plant variate, as in the rate of dry matter production per unit area of land.

V *Integral durations*. These are estimates of the areas beneath plots of primary or derived quantities and time, such as leaf area × time.

The following accounts identify many different examples of these five. In addition, we later see a number of very important interrelations between them, one frequent type being II = III × IV.

It is usual to *define* all of these quantities and relationships instantaneously, that is, as they stand at a single point in time. But although this method of definition provides an exact exposition, it was for many years necessary to perform *evaluations* of the various quantities in the form of mean values over a stated time interval – the method, in fact, of the classical approach to the subject.

If readers remain hazy over the mathematical implications of this difference, an analogy may be invoked. Two ways exist in which the speed of a motor car during a journey can be described. One is to take a speedometer reading. This sets a more or less instantaneous value ('damping' excepted) to the current speed of the vehicle which, of course, fluctuates widely during most journeys, often changing continuously. The other method is to ignore the actual fluctuations in speed experienced *en route*, calculating instead the mean value of speed over the whole journey from a knowledge of the total distance travelled and the time taken.

Mean values are thus generated by the classical approach and instantaneous values by the functional approach, though there are exceptions to this rule. In passing, we should note that only instantaneous values may properly be represented as single points on a progression plotted against time. Mean values should appear as a histogram, with a class interval equal to the harvest interval. This requirement is often neglected. Table 1.6 provides a synopsis of mathematical definitions.

**Table 1.6 Definitions and formulae** Instantaneous and mean values of each type of quantity are listed. Except in the case of Type I, the formulae for mean values are approximate.

| Type of quantity | Instantaneous definition | Formula for mean value over the interval $t_1$ to $t_2$ |
|---|---|---|
| I | $dY/dt$ | $(Y_2 - Y_1)/(t_2 - t_1)$ |
| II | $(1/Y)(dY/dt)$ | $(\log_e Y_2 - \log_e Y_1)/(t_2 - t_1)$ |
| III | $Y/Z$ | $((Y_1/Z_1) + (Y_2/Z_2))/2$ |
| IV | $(1/Z)(dY/dt)$ | $(Y_2 - Y_1)/(t_2 - t_1) \times (\log_e Z_2 - \log_e Z_1)/(Z_2 - Z_1)$ |
| V | $\int_{t_1}^{t_2} Y dt$ | $(Y_1 + Y_2)(t_2 - t_1)/2$ |

So, bearing in mind throughout that there are two chief levels of organization, five chief types of derived quantity and two methods of calculation, we can now move on to see a catalogue of the derived quantities.

# CHAPTER TWO
# *Absolute growth rates*

*Underlying concept*

The simplest index of plant growth; a rate of change in size, an increment in size per unit time. Most commonly applied to total dry weight or total leaf area per plant.

*Symbol and definition*

**G**, the rate of increase of total dry weight per plant, $W$.

*Dimensions and units*

$\mathrm{M\,T^{-1}}$; e.g. $\mathrm{g\,day^{-1}}$.

*Formulae*

Instantaneously, $\mathbf{G} = \mathrm{d}W/\mathrm{d}t$. The mean value over the interval $t_1$ to $t_2$ is given by

$$\bar{\mathbf{G}} = (W_2 - W_1)/(t_2 - t_1).$$

*Methods of calculation*

Instantaneously derived from functions fitted to $W$ versus $t$; if $W = \mathrm{f_W}(t)$, then $\mathbf{G} = \mathrm{f_W}'(t)$. Mean values are obtained from the separate destructive estimates $W_1$ and $W_2$ made at times $t_1$ and $t_2$ respectively.

*Relevant examples*

Mean values of absolute growth rate in Kreusler's maize (Fig. 2.1) and instantaneous values in sunflower and birch (Fig. 2.2).

For relations to other quantities see pp. 96–97.

For further details see CAUSTON & VENUS (p. 17); HUNT (p. 16).

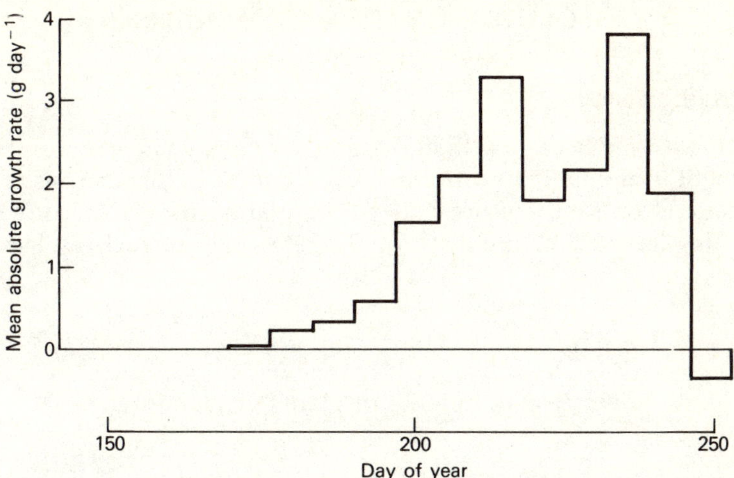


**Figure 2.1 Absolute growth rate in Kreusler's maize** Mean values, calculated for each of the harvest intervals shown in Table 1.1, are plotted here against time. The first four intervals produce values too small to plot on this scale and the rate during the final interval is negative.

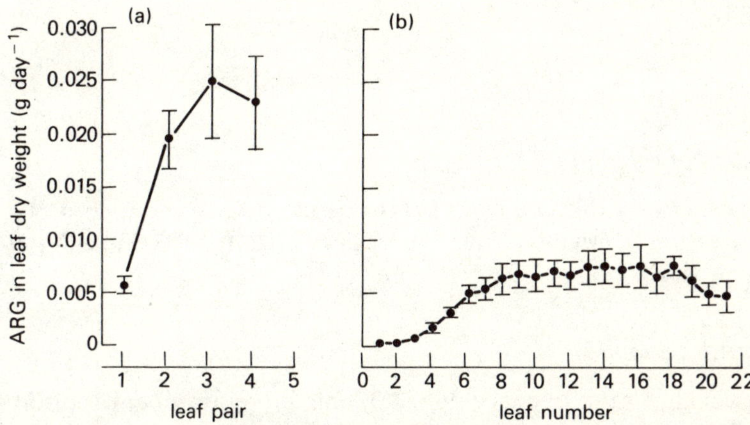


**Figure 2.2 Absolute growth rate in sunflower and birch** Instantaneous values derived from fitting Richards functions (see p. 104), (a) for sunflower and (b) for birch. Values are plotted against leaf developmental scales rather than against time (see Chapter 7). (From: D. R. Causton & J. C. Venus (1981). *The Biometry of Plant Growth*. London: Edward Arnold.)

*Underlying concept*

The simplest index of growth in number; a rate of change in number, an increment in number per unit time. Commonest in the analysis of plant populations or algal cultures; in whole plants, its use is confined to describing growth in numbers of discrete organs, such as leaves or roots.

*Symbol and definition*

**G**, the rate of increase in number of plants or plant organs, $N$.

*Dimensions and units*

$\mathrm{N\,T^{-1}}$; e.g. number $\mathrm{day}^{-1}$ or number $\mathrm{week}^{-1}$.

*Formulae*

Instantaneously, $\mathbf{G} = \mathrm{d}N/\mathrm{d}t$. The mean value over the interval $t_1$ to $t_2$ is given by

$$\bar{\mathbf{G}} = (N_2 - N_1)/(t_2 - t_1).$$

*Methods of calculation*

Instantaneously derived from functions fitted to $N$ versus $t$; if $N = f_N(t)$, then $\mathbf{G} = f_N'(t)$. Mean values are obtained from the counts $N_1$ and $N_2$ made at times $t_1$ and $t_2$ respectively.

*Relevant examples*

Classical algal growth curve (Fig. 2.3) and flower number in cotton crops (Fig. 2.4).

For further details see CAUSTON & VENUS (p. 17); HUNT (p. 16).

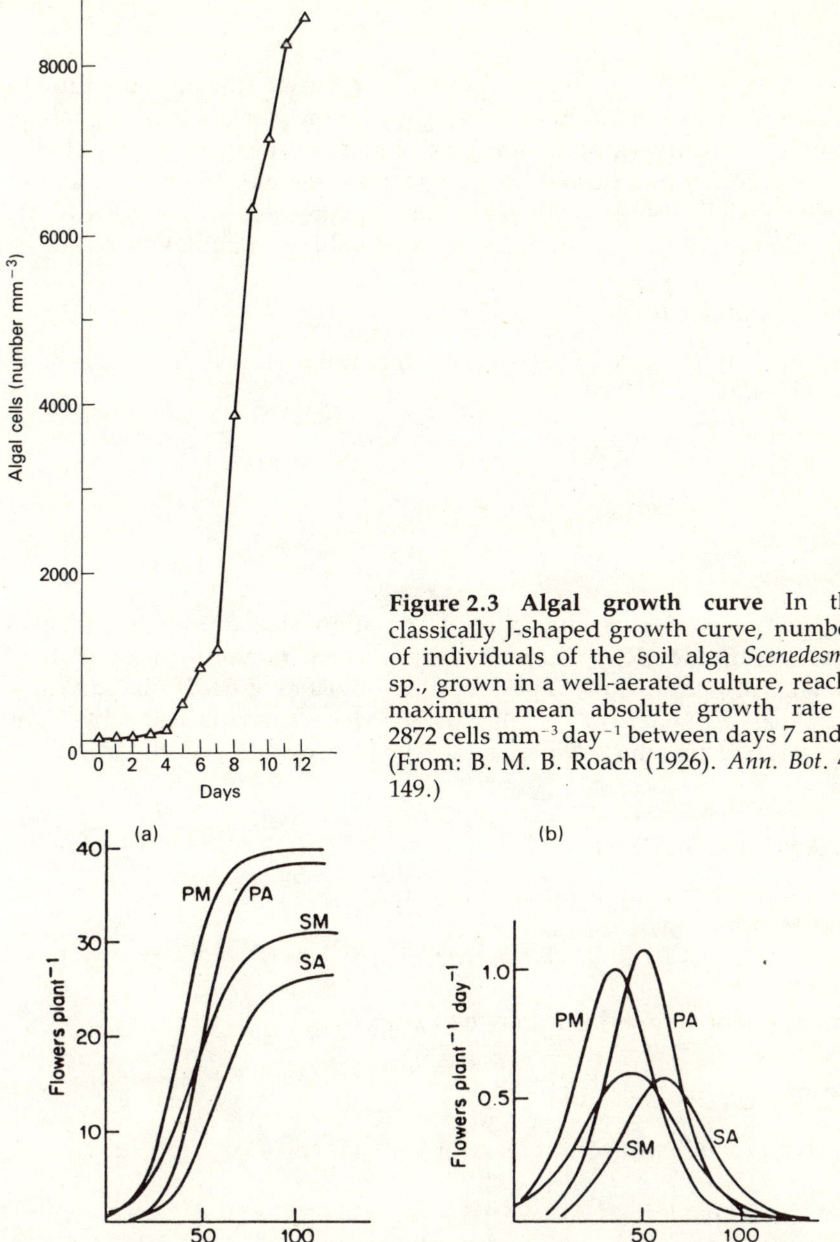


**Figure 2.3 Algal growth curve** In this classically J-shaped growth curve, numbers of individuals of the soil alga *Scenedesmus* sp., grown in a well-aerated culture, reach a maximum mean absolute growth rate of 2872 cells mm$^{-3}$ day$^{-1}$ between days 7 and 8. (From: B. M. B. Roach (1926). *Ann. Bot.* 40, 149.)

**Figure 2.4 Flowering in cotton** (a) Logistic growth curves (see p. 104) for cotton grown in Egypt in 1920; P. variety Pilion; S, variety Sakellaridis; M, sown in March; A, sown in April. (b) Instantaneous absolute growth rates in flower number, derived as the slopes of the curves given in (a). (From: J. A. Prescott (1922). *Ann Bot.* 36, 121.)

*Underlying concept*

During exponential (geometric) population growth, the absolute growth rate in number of individuals is controlled by a constant intrinsic rate of increase (a relative growth rate, q.v.), and by the current population size. If internal or external limiting factors operate, then sigmoidal or logistic growth results, with a gradual slowing of absolute growth rate towards zero at maximum or asymptotic value of population size.

*Symbol and definition*

(No symbol); the rate of increase of total number of individuals, $N$, in the population.

*Dimensions and units*

$\mathrm{N\,T^{-1}}$; e.g. number $\mathrm{day^{-1}}$.

*Formulae*

Instantaneous rate of change of population size during exponential growth is $\mathrm{d}N/\mathrm{d}t = \mathbf{R}N$, where $\mathbf{R}$ is the intrinsic rate of increase. Rate of change of population size during logistic growth is $\mathrm{d}N/\mathrm{d}t = \mathbf{R}N(k - N)/k$, where $k$ is the asymptotic 'carrying capacity', the maximum value which $N$ may take. The mean value over the interval $t_1$ to $t_2$ is $(N_2 - N_1)/(t_2 - t_1)$.

*Methods of calculation*

Derived from functions fitted to $N$ versus $t$;

$$\text{if } N = f_N(t), \text{ then } \mathrm{d}N/\mathrm{d}t = f_N'(t).$$

For exponential growth, it may be easier to fit

$$\log_e N = f_N(t),$$

in which case $\mathrm{d}N/\mathrm{d}t = f_N'(t) \times f_N(t)$.

*Relevant examples*

Growth and decay in buttercup populations (Table 2.1) and survivorship table for *Phlox* (Table 2.2).

For further details see SILVERTOWN (1982, p. 6).

**Table 2.1 Growth and decay in buttercups** Data for three *Ranunculus* species in a field in North Wales. Each set is taken from a plot where the species had 'moderate' density (From: J. Sarukhan & J. L. Harper (1973). *J.Ecol.* 61, 675.)

| Quantity | *R. acris* | *R. bulbosus* | *R. repens* |
|---|---|---|---|
| (a) Number of plants at start of observations ($m^{-2}$) | 62 | 90 | 117 |
| (b) Number of plants at end of observations, 2 years later ($m^{-2}$) | 124 | 64 | 139 |
| (c) Mean absolute growth rate ($m^{-2}$ $year^{-1}$) | 31 | −13 | 11 |

**Table 2.2 Survivorship in *Phlox drummondii*** Showing negative *absolute* growth rates for each interval. *See also* Table 3.2 (From: W. J. Leverich & D. A. Levin (1979). *Am. Nat.* 113, 881.)

| Age interval (days) | Length of interval (days) | No. present at beginning of interval | Mean absolute growth rate during interval ($day^{-1}$) |
|---|---|---|---|
| 0– 63 | 63 | 996 | −5.21 |
| 63–124 | 61 | 668 | −6.11 |
| 124–184 | 60 | 295 | −1.75 |
| 184–215 | 31 | 190 | −0.45 |
| 215–231 | 16 | 176 | −0.13 |
| 231–247 | 16 | 174 | −0.06 |
| 247–264 | 17 | 173 | −0.06 |
| 264–271 | 7 | 172 | −0.28 |
| 271–278 | 7 | 170 | −0.43 |
| 278–285 | 7 | 167 | −0.28 |
| 285–292 | 7 | 165 | −0.85 |
| 292–299 | 7 | 159 | −0.14 |
| 299–306 | 7 | 158 | −0.51 |
| 306–313 | 7 | 154 | −0.43 |
| 313–320 | 7 | 151 | −0.57 |
| 320–327 | 7 | 147 | −1.57 |
| 327–334 | 7 | 136 | −4.42 |
| 334–341 | 7 | 105 | −4.42 |
| 341–348 | 7 | 74 | −7.42 |
| 348–355 | 7 | 22 | −3.14 |
| 355 | – | 0 | – |

## CONCLUDING REMARKS

The Type I quantities are the simplest possible measures of plant growth rate.

Naturally, they are most useful when dealing with situations in which there is an interest in the absolute size of plants or plant populations, e.g. when parts of crops such as fruits or flowers need to be of a certain size before they become marketable, or when population size is approaching a fixed threshold or carrying capacity. Though absolute growth rates can be valuable comparative tools when they are used within bodies of like data (as in Fig. 2.4), when used to compare unlike systems (as between, say, Fig. 2.3 and Table 2.1) their usefulness declines.

To compare the overall performance of unlike systems, other approaches are needed. For example, suppose that two different species have been grown for a week in an experiment, and that at the beginning of the week the mean dry weights were 1 g per plant for the first species and 10 g per plant for the second. If at the end of the week the two mean dry weights were found to be 2 g and 11 g respectively, which species can we say had grown faster?

In one sense, the two performances were identical, equal weights having been gained over an equal period of time, in fact at a mean absolute growth rate of 1 g $week^{-1}$. But knowing that their initial dry weights were so dissimilar, it is easy to see that the performance of the first species, which doubled its dry weight, is in an important sense superior to that of the second, which increased its weight by only a tenth. Clearly, some measure of growth is needed which takes into account this original difference in size. That measure is the relative growth rate, described in the next chapter. Mean values of this rate for the two species in question would have been 0.69 and 0.10 $g\,g^{-1}\,week^{-1}$.

CHAPTER THREE

# *Relative growth rates*

## RELATIVE GROWTH RATE (RGR)

### *Underlying concept*

Called an 'efficiency index' by V. H. Blackman, RGR expresses growth in terms of a rate of increase in size per unit of size. This allows more equitable comparisons than an absolute growth rate. Normally, RGR deals with total dry weight, but other measures of size may be used (see next entry). In the financial world, RGR is analogous to the rate of compound interest earned on capital. Negative RGRs are relative *decay* rates.

### *Symbol and definition*

**R**, the rate of increase of total dry weight per plant, $W$, expressed per unit of $W$.

### *Dimensions and units*

$\mathrm{M\,M^{-1}T^{-1}}$; typically $\mathrm{g\,g^{-1}\,day^{-1}}$ or $\mathrm{g\,g^{-1}\,week^{-1}}$ (if both weights are in the same units they can be cancelled out); per cent per time may be used.

### *Formulae*

Instantaneously, $\mathbf{R} = (1/W)(\mathrm{d}W/\mathrm{d}t)$. This is the same as $\mathrm{d}(\log_e W)/\mathrm{d}t$. The mean value over the interval $t_1$ to $t_2$ is given by

$$\bar{\mathbf{R}} = (\log_e W_2 - \log_e W_1)/(t_2 - t_1).$$

### *Methods of calculation*

Instantaneously derived from functions fitted to $\log_e W$ versus $t$; if $\log_e W = f_W(t)$, then $\mathbf{R} = f_W'(t)$. Mean values are obtained from the separate destructive estimates $W_1$ and $W_2$ made at times $t_1$ and $t_2$ respectively.

### *Relevant examples*

Ontogenetic drift in relative growth rate in Kreusler's maize (Fig. 3.1) and variations in mean relative growth rate within four levels of organization (Fig. 3.2).

For relations to other quantities see pp. 88–91.

For further details see EVANS (p. 196); CAUSTON (p. 159); CAUSTON & VENUS (p. 17); HUNT (p. 16).

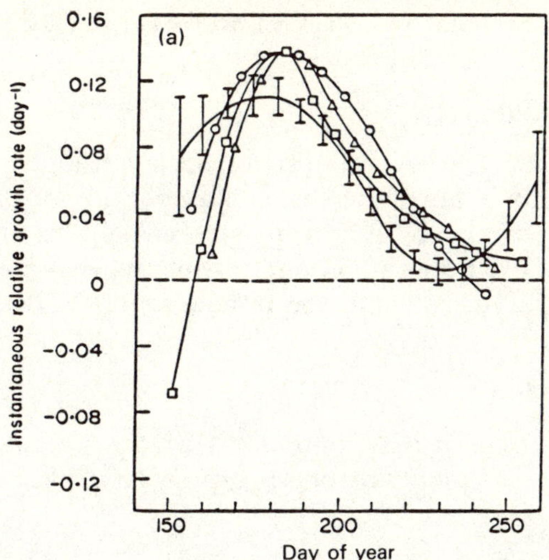


**Figure 3.1 Relative growth rates in Kreusler's maize** Instantaneous values derived from fitted spline curves (see p. 105). The curve bearing 95% limits is for maize grown in 1875. Three other years' data are also shown: ( ○ ) 1876, ( □ ) 1877 and ( △ ) 1878 (see also Table 1.1 and Fig. 8.2). (From: R. Hunt & G. C. Evans (1980). *New Phytol. 86*, 155.)

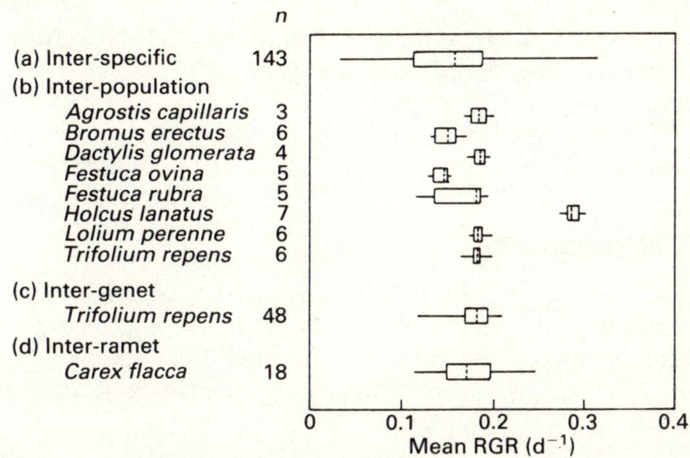


**Figure 3.2 Variation in mean relative growth rate** Four levels of organization are compared here. At each level, *n* gives the number of samples examined. The 'box and whisker' diagrams identify the ranges and quartiles of the RGR distribution within each sample. Twofold variation is common, except between species where it is tenfold. (From: R. Hunt (1987). *New Phytol.* 106 (Suppl.), 235.)

*Underlying concept*

Like RGR in total dry weight, other 'efficiency indices' may express growth in terms of a rate of increase in size per unit of size. The concept is again most useful where current size is held to influence current increase in size. Examples are Gregory's relative leaf [area] growth rate ($RL_AGR$), and the relative nutrient uptake rate (RNUR).

*Symbols and definitions*

$\mathbf{R}_L$ or $\mathbf{R}_M$, the rates of increase of total leaf area per plant or total mineral nutrient content per plant, $L_A$ or $M$, expressed per unit of $L_A$ or of $M$.

*Dimensions and units*

$A\,A^{-1}T^{-1}$ ($\mathbf{R}_L$) and $M\,M^{-1}T^{-1}$ ($\mathbf{R}_M$); e.g. $mm^2\,mm^{-2}\,day^{-1}$ and $mg\,mg^{-1}\,day^{-1}$ ($\mathbf{R}_M$), (if both areas, or both weights, are in the same units they cancel one another out); per cent per day may also be used.

*Formulae*

Instantaneously, $\mathbf{R}_L = (1/L_A)(dL_A/dT)$ and $\mathbf{R}_M = (1/M)(dM/dt)$. These are the same as $d(\log_e L_A)/dt$ and $d(\log_e M)/dt$. The mean values over the interval $t_1$ to $t_2$ are given by

$$\bar{\mathbf{R}}_L = (\log_e L_{A2} - \log_e L_{A1})/(t_2 - t_1); \quad \bar{\mathbf{R}}_M = (\log_e M_2 - \log_e M_1)/(t_2 - t_1).$$

*Methods of calculation*

Instantaneously derived from functions fitted to $\log_e L_A$, or to $\log_e N$, versus $t$. If $\log_e L_A = f_L(t)$, then $\mathbf{R}_L = f_L'(t)$; if $\log_e N = f_N(t)$, then $\mathbf{R}_N = f_N'(t)$. Mean values are obtained from the separate destructive estimates $L_{A1}$ and $L_{A2}$, or $N_1$ and $N_2$; these are made at times $t_1$ and $t_2$ respectively.

*Relevant examples*

Relative leaf area growth rate in Kreusler's maize (Fig. 3.3) and relative nitrogen uptake rate in woody saplings (Table 3.1)

For further details see EVANS (p. 357); CAUSTON & VENUS (p. 17); HUNT (p. 22).

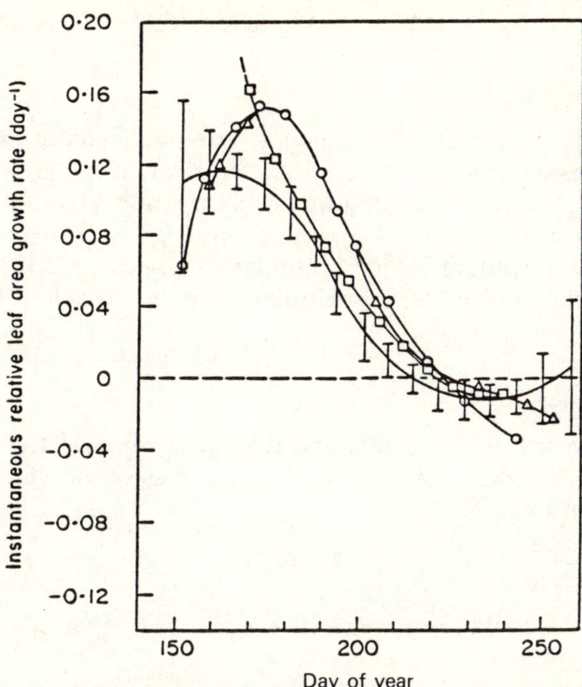


**Figure 3.3 Relative leaf area growth rates in Kreusler's maize** Instantaneous values derived from fitted spline curves (see p. 105). The curve bearing 95% limits is for maize grown in 1875. three other years' data are also shown: ( ○ ) 1876, ( □ ) 1877 and ( △ ) 1878 (see also Table 1.1). (From: R. Hunt & G. C. Evans (1980). *New Phytol.* 86, 155.)

**Table 3.1 Relative nutrient uptake rate in woody saplings** Results for RGR in nitrogen content, $\mathbf{R}_N$, during a short exponential phase of growth at varying levels of external nitrogen supply. (From: T. Ingestad (1981). *Physiologia Pl.* 52, 545.)

| Species | Treatment (% increase day$^{-1}$ in external nitrogen | $R_N$* (% day$^{-1}$) |
|---|---|---|
| Birch | 5 | 4.4 |
| | 10 | 9.2 |
| | 15 | 14.6 |
| | 25 | 25.8 |
| Grey alder | 5 | 6.0 |
| | 10 | 11.0 |
| | 15 | 16.2 |
| | 19 | 18.4 |

*each treatment exhibited a constant internal N concentration, therefore $R_N$ is also equal to $R_W$ (RGR in total dry weight)

*Underlying concept*

In exponential (geometric) population growth, increase in number of individuals is controlled by a constant 'intrinsic rate', an RGR, and by current population size. If limiting factors operate, sigmoidal or logistic growth can result, with a gradual slowing of growth towards a maximum (asymptotic) value of population size; here, the intrinsic rate emerges only at the very beginning and a variable RGR is seen elsewhere.

*Symbol and definition*

$r$ is the intrinsic rate of increase and $\mathbf{R}$ is the current relative growth rate in total number, $N$, both expressed as number of individuals per unit number of individuals.

*Dimensions and units*

$N\,N^{-1}\mathrm{T}^{-1}$; the number dimensions cancel out to give, e.g., $\mathrm{day}^{-1}$.

*Formulae*

Instantaneously during exponential growth, $\mathrm{d}N/\mathrm{d}t = \mathbf{R}N$, where $\mathbf{R}$ is both the intrinsic rate of increase and the current RGR. Hence, $\mathbf{R} = (1/N)(\mathrm{d}N/\mathrm{d}t)$. During logistic growth, $\mathrm{d}N/t = rN(k - N)/k$, where $r$ is the intrinsic or initial rate of increase and $k$ is the asymptotic 'carrying capacity', the maximum value which $N$ may take. Here, $\mathbf{R} = r(k - N)/k$ and $r = \mathbf{R}k(k - N)$. A mean value over the interval $t_1$ or $t_2$ is given by

$$\bar{\mathbf{R}} = (\log_e N_2 - \log_e N_1)/(t_2 - t_1).$$

*Methods of calculation*

For exponential growth, derive the function $\log_e N = a + bt$, where $a$ and $b$ are constants; then $b = (1/N)(\mathrm{d}N/\mathrm{d}t) = \mathbf{R} = r$. For logistic growth, fit $N$ as a logistic function of $t$; $r$ emerges as the 'slope' parameter of the regression and the formula given above may be used to derive $\mathbf{R}$.

*Relevant examples*

Growth in duckweed (Fig. 3.4) and decay in *Phlox* population (Table 3.2).

For further details see SILVERTOWN (p. 6).

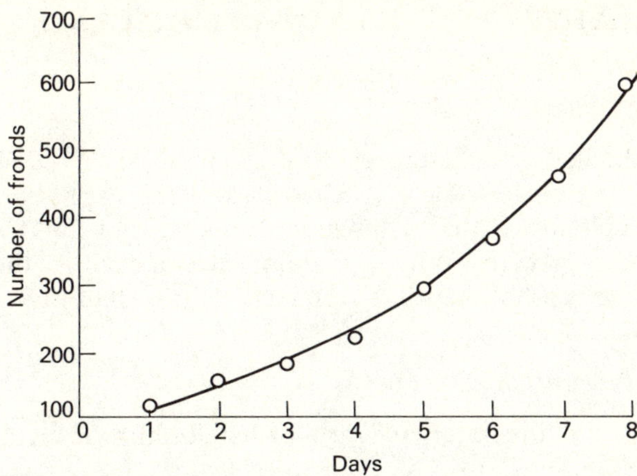


**Figure 3.4 Exponential growth in duckweed** Observed and fitted data for a colony grown under continuous illumination. The equation of the fitted curve is $N = 90.5e^{0.316t}$, implying an intrinsic rate of population increase of 0.316 day$^{-1}$. (From: E. Ashby (1929). *Ann. Bot.* 43, 333.)

**Table 3.2 Survivorship in *Phlox drummondii*** Showing negative *relative* growth rates for each interval. *See also* Table 2.2. (From: W. J. Leverich & D. A. Levin (1979). *Am.Nat.* 113, 881.

| Age interval (days) | Length of interval (days) | No. present at beginning of interval | Mean *relative* growth rate during interval (day$^{-1}$) |
|---|---|---|---|
| 0– 63 | 63 | 996 | −0.0063 |
| 63–124 | 61 | 668 | −0.0133 |
| 124–184 | 60 | 295 | −0.0073 |
| 184–215 | 31 | 190 | −0.0025 |
| 215–231 | 16 | 176 | −0.0007 |
| 231–247 | 16 | 174 | −0.0004 |
| 247–264 | 17 | 173 | −0.0003 |
| 264–271 | 7 | 172 | −0.0017 |
| 271–278 | 7 | 170 | −0.0025 |
| 278–285 | 7 | 167 | −0.0017 |
| 285–292 | 7 | 165 | −0.0053 |
| 292–299 | 7 | 159 | −0.0009 |
| 299–306 | 7 | 158 | −0.0037 |
| 306–313 | 7 | 154 | −0.0028 |
| 313–320 | 7 | 151 | −0.0038 |
| 320–327 | 7 | 147 | −0.0111 |
| 327–334 | 7 | 136 | −0.0370 |
| 334–341 | 7 | 105 | −0.0500 |
| 341–348 | 7 | 74 | −0.1732 |
| 348–355 | 7 | 22 | – |
| 355 | – | 0 | – |

*Underlying concept*

In microbial, algal or cell cultures, RGRs expressed on a daily or weekly basis can be high (see Table 1.2). Growth is better expressed as the time taken to double in weight (or whatever): the shorter the doubling time, the faster the growth. Where systems are decaying, the (negative) doubling time is a halving time, referred to as a 'half life'.

*Symbol and definition*

(No symbol); the time interval over which a doubling (halving) of weight (etc.) occurs.

*Dimensions and units*

T; e.g. second, minute, hour.

*Formulae*

Instantaneous doubling (or halving) time is 0.693/**R**, where **R** is instantaneous relative growth (or decay) rate. The mean value over the interval $t_1$ to $t_2$ is 0.693/**R**, or $0.693 \times (t_2 - t_1)/(\log_e W_2 - \log_e W_1)$, where $W$ may be any measure of size. The constant 0.693 is the value of $\log_e 2$.

*Methods of calculation*

From the above instantaneous or mean-value formulae. Instantaneous **R** may be derived from functions fitted to $\log_e W$ versus $t$; if $\log_e W = f_W(t)$, then $\mathbf{R} = f_W'(t)$.

*Relevant examples*

Batch culture dynamics (Fig. 3.5) and continuous culture dynamics (Fig. 3.6).

For further details see COOMBS & HALL (1982, p. 66); HUNT (p. 22); H. G. Schlegel (1986). *General Microbiology*. Cambridge: Cambridge University Press.

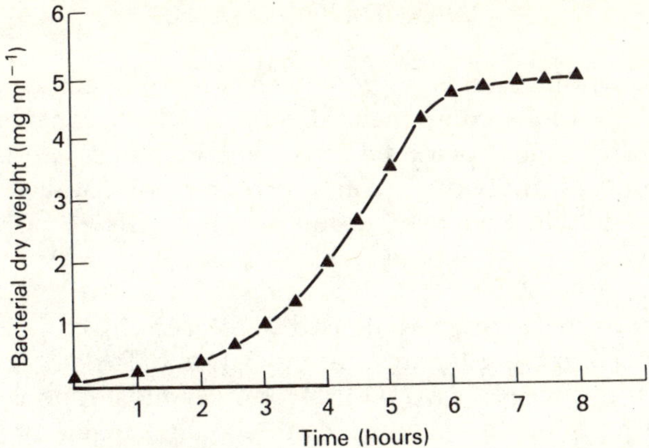


**Figure 3.5 Batch culture dynamics** A classically sigmoidal growth curve obtained for the virulent V8 strain of *Staphylococcus aureus* grown in a closed culture system. The intrinsic rate of increase is approximately 0.92 hour$^{-1}$, the initial doubling time approximately 0.76 hours, and the asymptotic carrying capacity of the medium approximately 5.1 mg ml$^{-1}$. (From: S. Arvidson *et al.* (1976). In: A. C. R. Dean *et al.* (eds), *Continuous Culture, Volume 6*. Chichester: Ellis Horwood.)

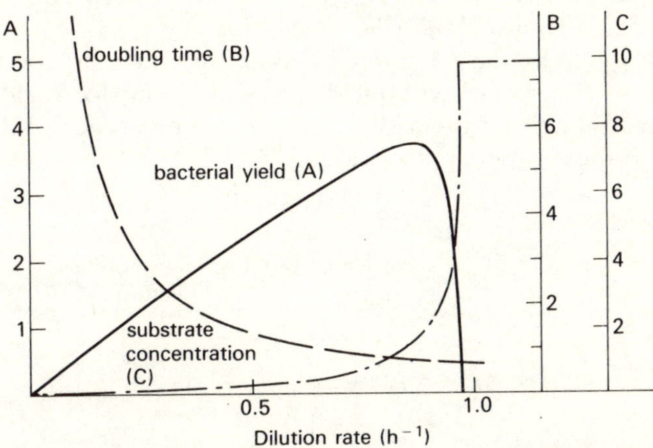


**Figure 3.6 Continuous culture dynamics** Predicted relations between bacterial yield (A, g l$^{-1}$ hour$^{-1}$), doubling time (B, hours), and substrate concentration (C, g l$^{-1}$) in a continuously diluted chemostat vessel containing a bacterium with an intrinsic rate of increase of 1.0 hour$^{-1}$, a substrate requirement of 0.2 g l$^{-1}$ at half this maximum rate, and supplied with an incoming substrate concentration of 10 g l$^{-1}$. (From: D. Herbert *et al.* (1956). *J. gen. Microbiol.* 14, 601.)

As Type II quantities, relative growth rates provide convenient integrations of the various component processes which contribute to the performances of whole plant systems, whether these are themselves plant components, individual plants, or plant populations. RGRs are especially useful when we need to compare the performances of unlike material (e.g. on p. 24 and in Table 1.2 and Fig. 3.2) or of unlike treatments.

The comparison between RGR and AGR begun on p. 24 may be continued by comparing Figures 2.1 and 3.1, and Tables 2.2 and 3.2.

RGR is a pure concept, say $(1/Y)(\mathrm{d}Y/\mathrm{d}t)$, depends upon the assumption that all parts of the quantity $Y$ are equally capable of producing further amounts of $Y$, in the same way that invested capital accumulates through the payment of compound interest. If this were so in the case of total dry weight, the plant would grow exponentially and RGR would be constant for material of any one genetic provenance and environmental regime. However, we know that as most plants grow the proportion of purely structural material which they contain increases (Figs. 4.1 and 4.5). This happens for much the same reason that larger animals develop proportionally more bulky bones than smaller ones. So, RGR soon declines with time (Fig. 3.1) and the interest then passes to the components of RGR which together explain exactly how this ontogenetic decline comes about (Fig. 8.2).

Perhaps it is in the case of populations of unicellular organisms, which can increase in number without division of labour for so long as the environment supports them (Fig. 3.4), that the concept of RGR as an 'efficiency index' is best preserved.

# CHAPTER FOUR

# *Simple ratios*

## LEAF AREA RATIO (LAR)

### *Underlying concept*

A morphological index of the leafiness of the plant, devised by G. E. Briggs and co-workers. A measure of the 'balance of payments' between 'income' and 'expenditure' because it deals with the potentially photosynthesizing and the potentially respiring components of the plant.

### *Symbol and definition*

**F**; the ratio between total leaf area per plant, $L_A$, and total dry weight per plant, $W$. More strictly a leaf area *quotient*, but the term 'ratio' is almost universally used.

### *Dimensions and units*

$A\,M^{-1}$; typically $mm^2\,mg^{-1}$ or $m^2\,g^{-1}$.

### *Formulae*

Instantaneously, $\mathbf{F} = L_A/W$. The mean value over the interval $t_1$ to $t_2$ is given by

$$\bar{\mathbf{F}} = ([L_{A1}/W_1] + [L_{A2}/W_2])/2 = (F_1 + F_2)/2.$$

### *Methods of calculation*

Instantaneously derived from functions fitted to $\log_e L_A$ and to $\log_e W$ versus $t$; if $\log_e L_A = f_L(t)$ and $\log_e W = f_W(t)$, then $\mathbf{F} = L_A/W = \exp[f_L(t) - f_W(t)]$. Unsmoothed instantaneous values are also available directly from the defining formula if $L_{A1}$ and $W_1$ are known; harvest interval means are available if $L_{A2}$ and $W_2$ are also known.

### *Relevant examples*

Leaf area ratio in Kreusler's maize (Fig. 4.1) and leaf area ratio in perennial ryegrass (Fig. 4.2).

For relations to other quantities see pp. 90–91.

For further details see EVANS (p. 212); CAUSTON & VENUS (p. 21); HUNT (p. 23).

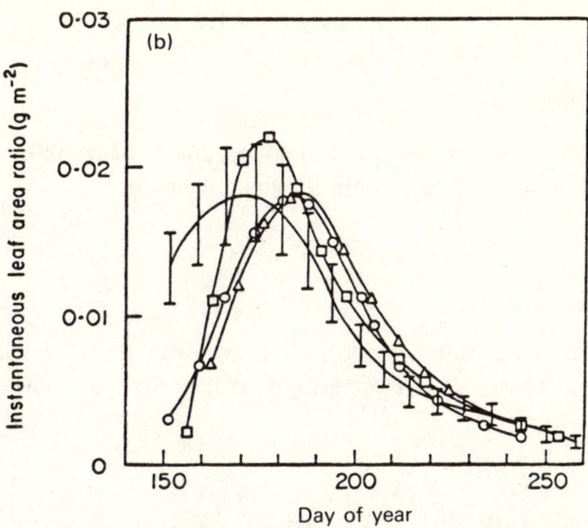


**Figure 4.1 Leaf area ratios in Kreusler's maize** Instantaneous values derived from fitted spline curves (see p. 105). The curve bearing 95% limits is for maize grown in 1875. Three other years' data are also shown: ( ○ ) 1876, ( □ ) 1877 and ( △ ) 1878 (see also Table 1.1 and Fig. 8.2). (From: R. Hunt & G. C. Evans (1980). *New Phytol*. 86, 155.)

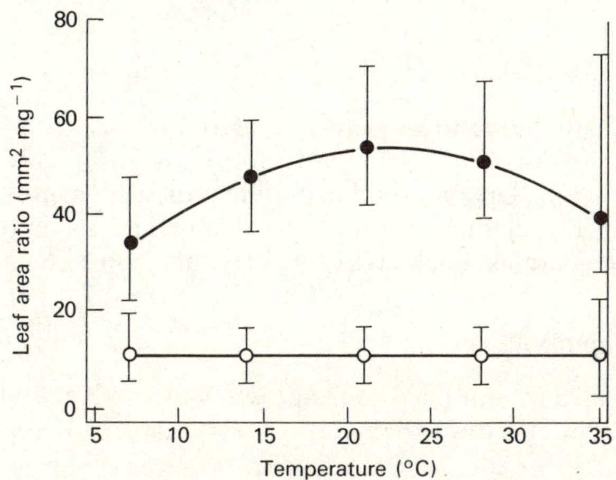


**Figure 4.2 Leaf area ratio in perennial ryegrass** Fitted values after 28 days' growth in a factorial experiment investigating temperature and light intensity. Open symbols are for material grown at full light intensity, where LAR is almost independent of temperature; closed symbols are for material grown at 20% of full light, where LAR is both raised overall and subject to a maximum of 23 °C. (R. Hunt *et al.*, unpubl.)

*Underlying concept*

An index of the leafiness of the leaf. A measure of density or of relative thinness, because it deals with leaves' areas in relation to their dry weight.

*Symbol and definition*

No symbol; the ratio between total leaf area per plant, $L_A$, and total leaf dry weight per plant, $L_W$. The term 'specific' means divided by weight.

*Dimensions and units*

$\mathrm{A\,M^{-1}}$; typically $\mathrm{mm^2\,mg^{-1}}$ or $\mathrm{m^2\,g^{-1}}$.

*Formulae*

Instantaneously, $L_A/L_W$. The mean value over the interval $t_1$ to $t_2$ is given by

$$([L_{A1}/L_{W1}] + [L_{A2}/L_{W2}])/2.$$

*Methods of calculation*

Instantaneously derived from functions fitted to $\log_e L_A$ and to $\log_e L_W$ versus $t$; if $\log_e L_A = f_A(t)$ and $\log_e L_W = f_W(t)$, then $L_A/L_W = \exp[f_A(t) - f_W(t)]$. Unsmoothed instantaneous values are also available directly from the defining formula if $L_{A1}$ and $L_{W1}$ are known; harvest interval means are available if $L_{A2}$ and $L_{W2}$ are also known.

*Relevant examples*

Specific leaf area in four glasshouse-grown species (Fig. 4.3) and specific leaf area in winter-lettuce (Fig. 4.4).

For relations to other quantities see pp. 90–91.

For further details see EVANS (pp. 215, 316); CAUSTON & VENUS (p. 22); HUNT (p. 26).

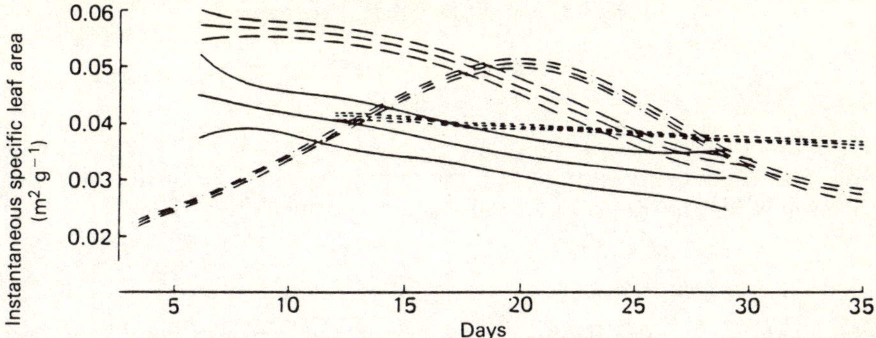


**Figure 4.3 Specific leaf area in four glasshouse-grown species** Instantaneous values derived from fitted Richards functions (see p. 104): wheat (solid lines), maize (dashed lines), sunflower (chain-dotted lines) and birch (dotted lines). The curves bear 95% limits throughout. (From: D. R. Causton & J. C. Venus (1981). *The Biometry of Plant Growth*. London: Edward Arnold.)

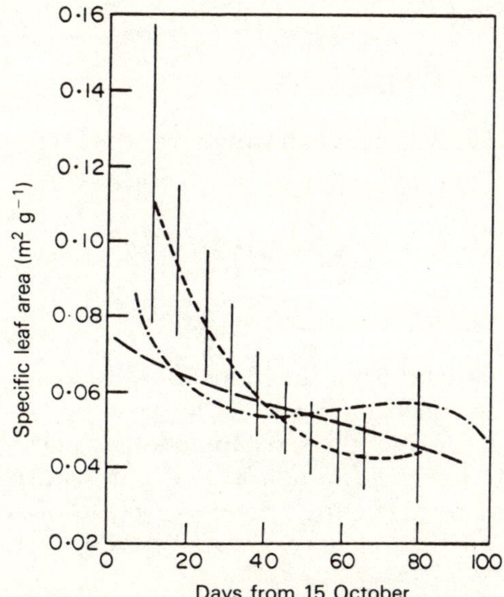


**Figure 4.4 Specific leaf area in winter lettuce** Crops planted under glass in mid-October, grown under $CO_2$ enrichment, and harvested in January. Three years' data are shown: 1978–79 (dotted line, with 95% limits), 1979–80 (chain-dotted line), 1980–81 (dashed line). See also Fig. 8.5. (From: Hunt *et al.* (1984). *Ann. Bot.* 54, 743.)

*Underlying concept*

An index of leafiness of the plant on a dry weight basis. A measure of the 'productive investment' of the plant, because it deals with the relative expenditure on potentially photosynthesizing organs.

*Symbol and definition*

No symbol; the ratio between total leaf dry weight per plant, $L_W$, and total dry weight per plant, $W$. More strictly a leaf weight fraction, but the term 'ratio' is widely used.

*Dimensions and units*

$M M^{-1}$, which is effectively dimensionless; units are numbers ($x$) within the range $0 < x > 1$.

*Formulae*

Instantaneously, $L_W/W$. The mean value over the interval $t_1$ to $t_2$ is given by

$$([L_{W1}/W_1] + [L_{W2}/W_2])/2.$$

*Methods of calculation*

Instantaneously derived from functions fitted to $\log_e L_W$ and to $\log_e W$ versus $t$; if $\log_e L_W = f_L(t)$ and $\log_e W = f_W(t)$, then $L_W/W = \exp[f_L(t) - f_W(t)]$. Unsmoothed instantaneous values are also available directly from the defining formula if $L_{W1}$ and $W_1$ are known; harvest interval means are available if $L_{W2}$ and $W_2$ are also known.

*Relevant examples*

Leaf weight ratio in four glasshouse grown species (Fig. 4.5) and leaf weight ratio in ryegrass and clover (Table 4.1).

For relations to other quantities see pp. 86–87.

For further details see EVANS (p. 215, 297); CAUSTON & VENUS (p. 22); HUNT (p. 27).

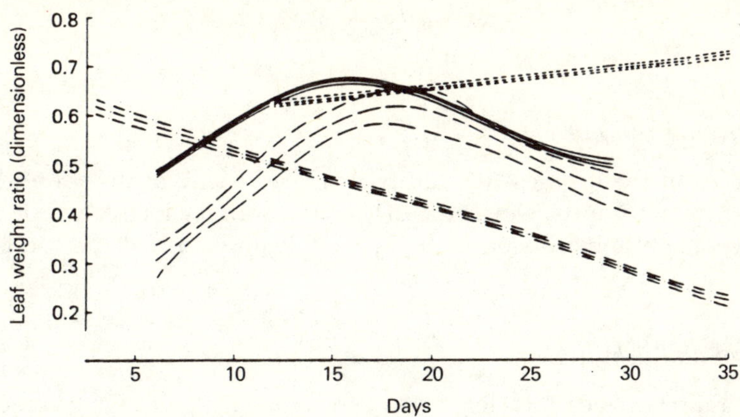


**Figure 4.5 Leaf weight ratio in four glasshouse-grown species** Instantaneous values derived from fitted Richards functions (see p. 104): wheat (solid lines), maize (dashed lines), sunflower (chain-dotted lines) and birch (dotted lines). The curves bear 95% limits throughout. (From: D. R. Causton & J. C. Venus (1981). *The Biometry of Plant Growth*. London: Edward Arnold.)

**Table 4.1 Leaf weight ratio in ryegrass and clover** Sixteen cultivars were examined over a three-week period. 95% confidence intervals are given in parenthesis. (From: C. O. Elias & M. J. Chadwick (1979). *J.Ecol*. 16, 537.)

| Species | Cultivar | Leaf weight ratio |
|---|---|---|
| *Lolium perenne* | S23 | 0.505 (0.023) |
| | S24 | 0.500 (0.017) |
| | Melle | 0.485 (0.022) |
| | Pelo | 0.512 (0.010) |
| | Standion | 0.512 (0.017) |
| | Endura | 0.497 (0.024) |
| *L. multiflorum* | Tiara | 0.547 (0.014) |
| *L. perenne* × *L. multiflorum* | Sabrina | 0.534 (0.029) |
| *Trifolium repens* | S100 | 0.452 (0.036) |
| | Hiua | 0.458 (0.011) |
| | S184 | 0.483 (0.018) |
| | Blanca | 0.471 (0.019) |
| | Kersey | 0.444 (0.032) |
| | Pajbjerg | 0.457 (0.039) |
| *Trifolium pratense* | Tetri | 0.467 (0.016) |
| *Trifolium hybridum* | ('British') | 0.167 (0.015) |

*Underlying concept*

An index of the leafiness of a crop, or more strictly of the ground area upon which it stands. Devised by D. A. Watson, it effectively represents the average number of complete layers of leaf material displayed by the crop.

*Symbol and definition*

L; the ratio between total leaf area of the crop, $L_A$, and total ground area upon which it stands, $P$; the leaf area per unit ground area.

*Dimensions and units*

$A\,A^{-1}$, which is dimensionless; values are positive pure numbers.

*Formulae*

Instantaneously, $\mathbf{L} = L_A/P$. The mean value over the interval $t_1$ to $t_2$ is given by

$$\bar{\mathbf{L}} = ([L_{A1}/P_1] + [L_{A2}/P_2])/2.$$

Most commonly $P_1 = P_2$, so the formula is simplified.

*Method of calculation*

Instantaneously derived from functions fitted to $L_A/P$ versus $t$, such as $L_A/P = f_L(t)$. The procedure is more complex for log-transformed data (HUNT, p. 57). Unsmoothed instantaneous values are also available directly from the defining formula if $L_{A1}$ and $P_1$ are known; harvest interval means are available if $L_{A2}$ and $P_2$ are also known.

*Relevant examples*

Seasonal changes in leaf area index (Fig. 4.6) and leaf area index in sugar cane (Fig. 4.7).

From relations to other quantities see pp. 92–93.

For further details see EVANS (p. 217); HUNT (p. 33).

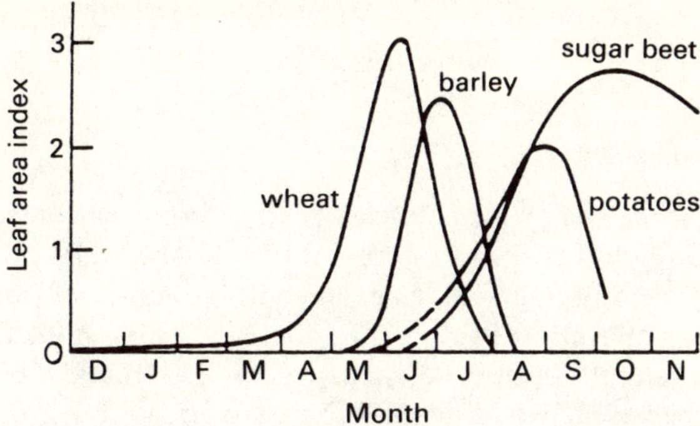


**Figure 4.6 Seasonal changes in leaf area index** Smoothed curves for four crops grown at Rothamsted Experimental Station in the 1940s. Modern crops are more leafy, but trends in LAI remain equally dependent upon the ontogeny of the crop rather than upon the time of year. (From: D. J. Watson (1947). *Ann. Bot.* 11, 41.)

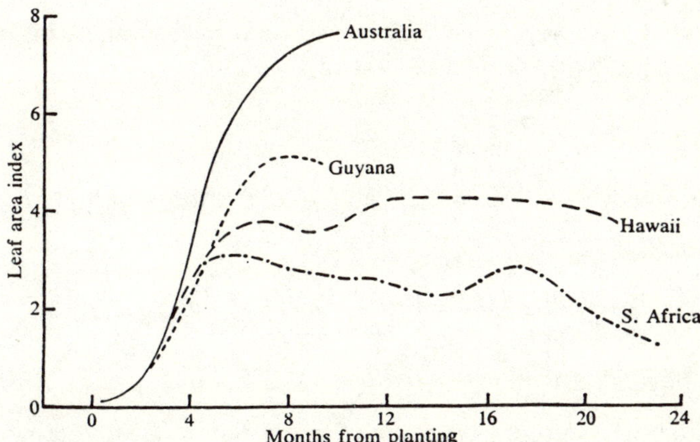


**Figure 4.7 Leaf area index in sugar cane** The development of LAI in crops grown under four contrasted regimes. (From: T. A. Bull & K. T. Glasziou (1975). In: L. T. Evans (ed.), *Crop Physiology: Some Case Histories*. Cambridge: Cambridge University Press.)

*Underlying concept*

A measure of the efficiency with which green plants convert solar energy, or solar energy plus energy input by humans, to chemically stored energy. Several efficiencies exist, with several possibilities on the input side and several on the output side (see Table 4.2). One combination gives digestible energy, a measure of the fraction of the energy fixed by the crop which is digestible by man or by livestock.

*Symbol and definition*

No symbol; ultimately, the ratio (quotient) between the energy output from a system, $I_2$, and the energy input $I_1$.

*Dimensions and units*

A dimensionless ratio, often expressed as a percentage; original units are usually joules ($J = 2.39 \times 10^{-4}$ kCal).

*Formulae*

Instantaneously $I_2/I_1$. The mean value over the interval $t_{(1)}$ to $t_{(2)}$ is given by

$$([I_{2(1)}/I_{1(1)}] + [I_{2(2)}/I_{1(2)}])/2.$$

*Method of calculation*

Instantaneously derived from functions fitted to $I_1$ and $I_2$ versus $t$; if $I_1 = f_1(t)$ and $I_2 = f_2(t)$, then $I_2/I_1 = f_2(t)/f_1(t)$. Unsmoothed instantaneous values are also available directly from the defining formula if $I_{1(1)}$ and $I_{2(1)}$ are known; harvest interval means are available if $I_{1(2)}$ and $I_{2(2)}$ are also known.

*Relevant examples*

Crop/environment efficiencies (Table 4.3) and crop/management efficiencies (Table 4.4). See also Table 5.1.

For further details see J. Phillipson (1966). *Ecological Energetics*. London: Edward Arnold.

**Table 4.2 Efficiencies of energy conversion** Numerators and/or denominators available for the calculation of efficiency ratios. The series reduces progressively in magnitude of energy content.

---

Total solar radiation incident upon the plant environment
Photosynthetically active radiation (PAR) incident upon the plant environment
PAR intercepted by the plant
PAR entering the plant
Energy fixed by the plant
Energy retained by the plant
Energy retained by the plant in 'desired' form or place

---

**Table 4.3 Crop/environment photosynthetic efficiencies** The percentage of net production in relation to visible light energy received in four crops. (From: PHILLIPSON, p. 35.)

| Crop | Locality | Days in leaf | Efficiency (%) |
|---|---|---|---|
| Scots pine | Britain | 360 | 2.2 to 2.6 |
| Sugar cane | Java | 360 | 1.9 |
| European beech | Denmark | 164 | 2.5 |
| Rice | Japan | 150 | 2.2 |

**Table 4.4 Crop/management efficiencies** The ratio of consumable output energy to consumed human, draught animal and fossil fuel input energy in several cropping systems. (From: L. T. Evans (ed.) (1975). *Crop Physiology: Some Case Histories*. Cambridge: Cambridge University Press.)

| | | |
|---|---|---|
| Fishing | Pacific atoll | 6 |
| Food gathering | Wild wheats, Turkey | 40–50 |
| Hand cultivation | 13 examples | 17 (3–34) |
| | Sweet potatoes, New Guinea | 16 |
| Draught animals | Two examples | 10 (6–14) |
| Mechanized agriculture | (a) Primary inputs only | 22 (9–34) |
| | (b) Including secondary inputs | |
| | Australia | 3 |
| | USA | 0.2 |

*Underlying concept*

A measure of the fraction of the crop which forms the payload or marketable component. Synonymous with the term 'harvest index'.

*Symbol and definition*

No symbol; the ratio (quotient) between the dry weight of the marketable component of the crop, $W_M$, and its total dry weight, $W$. (A fresh weight basis is also commonly used, *mutatis mutandis*.)

*Dimensions and units*

$MM^{-1}$, i.e. a dimensionless ratio, or fraction; often expressed as a percentage.

*Formulae*

Instantaneously, $W_M/W$; The mean value over the interval $t_1$ to $t_2$ is given by

$$([W_{M1}/W_1] + [W_{M2}/W_2])/2.$$

*Method of calculation*

During the development of $W_M$ in the crop, the index may be derived instantaneously from functions fitted to $\log_e W_M$ and $\log_e W$ versus $t$; if $\log_e W_M = f_M(t)$ and $W = f(t)$, then $W_M/W = \exp[f_M(t) - f(t)]$. Unsmoothed instantaneous values, say at a final harvest, are available directly from the defining formula; harvest interval means are available if additional, previous values of $W_M$ and $W$ are also known.

*Relevant examples*

Harvest index in woody and herbaceous crops (Table 4.5) and development of marketable components (Fig. 4.8).

For further details see the references cited in the examples.

**Table 4.5 Harvest index** Data for some forest and $C_3$ agricultural crops in temperature climates. (From: C. B. Johnson (ed.) (1981). *Physiological Processes Limiting Plant Productivity*. London: Butterworth.)

| Crop | Crop growth rate (q.v.) (mg $ha^{-1}$ $year^{-1}$) | Harvest index |
|---|---|---|
| Coniferous | | |
| *Cryptomeria japonica* | 53 | 0.65 |
| *Pinus radiata* | 46 | 0.66 |
| *Tsuga heterophylla* | 43 | 0.65 |
| *Abies sachaliensis* | 29 | 0.65 |
| *Pseudotsuga menziesii* | 28 | 0.71 |
| *Pinus nigra* | 25 | 0.46 |
| *Picea abies* | 22 | 0.61 |
| *Thuja plicata* | 20 | 0.68 |
| Agricultural | | |
| Sugar beet | 42 | 0.45 |
| Wheat | 30 | 0.40 |
| Perennial ryegrass | 26 | 0.85 |
| Potatoes | 22 | 0.82 |
| Field bean | 20 | 0.31 |
| Barley | 18 | 0.39 |
| Calabrese | 12 | 0.17 |
| Soya beans | 10 | 0.30 |

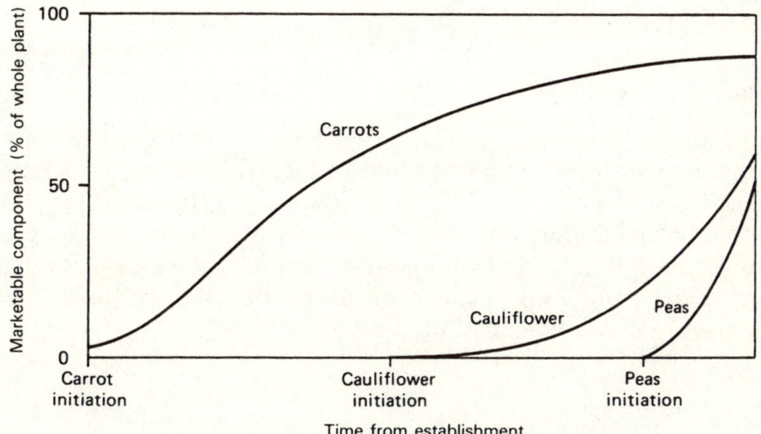


**Figure 4.8 Development of marketable components** In some crops the marketable component establishes early in the growing period, but in others it does not. (From: J. K. A. Bleasdale (1981). In: C. R. W. Spedding (ed.), *Vegetable Productivity*. London: Macmillan.)

*Underlying concept*

An index of the proportion of protein in the total crop or in the marketable part of the crop.

*Symbol and definition*

No symbol; the weight of protein, $W_{pr}$, per unit of marketable dry weight, $W_M$, or per unit of total dry weight, $W$. (A fresh weight basis is also commonly used, *mutatis mutandis*.)

*Dimensions and units*

$MM^{-1}$, i.e. a dimensionless ratio, or fraction; often expressed as a percentage.

*Formulae*

Instantaneously, $W_{pr}/W$; The mean value over the interval $t_1$ to $t_2$ is given by

$$([W_{pr1}/W_1] + [W_{pr2}/W_2])/2,$$

where $W$ can be substituted by $W_M$.

*Method of calculation*

During the development of $W_{pr}$ in the crop, the index may be derived instantaneously from functions fitted to $\log_e W_{pr}$ and $\log_e W$ versus $t$; if $\log_e W_{pr} = f_{pr}(t)$ and $\log_e W = f(t)$, then $W_{pr}/W = \exp[f_{pr(t)} - f(t)]$. Unsmoothed instantaneous values, say at a final harvest, are available directly from the defining formula; harvest interval means are available if additional, previous values of $W_{pr}$ and $W$ are also known. Throughout, $W$ can be substituted by $W_M$.

*Relevant examples*

World production of edible dry matter and protein (Table 4.6).

For further details see the references cited in the example.

**Table 4.6 World production of edible dry matter and protein** Figures from FAO, UN and other sources. (From: L. T. Evans (ed.) (1975). *Crop Physiology: Some Case Histories*. Cambridge: Cambridge University Press.)

| Crop | Edible dry matter (metric tons × $10^7$) | | Protein (metric tons × $10^6$) | | Percent protein | |
|---|---|---|---|---|---|---|
| Cereal grains | | | | | | |
| Wheat | 27.5 | | 32.9 | | 11.9 | |
| Rice | 26.7 | | 23.2 | | 8.7 | |
| Maize | 23.5 | | 24.7 | | 10.5 | |
| Barley | 11.4 | | 11.6 | | 10.2 | |
| Sorghum/millet | 8.2 | | 7.4 | | 9.0 | |
| Others | 7.6 | | 1.1 | | 1.4 | |
| | ___ | 104.9 | ___ | 100.9 | ___ | 9.6 |
| Starchy roots | | | | | | |
| Potato | 6.6 | | 6.0 | | 9.1 | |
| Sweet potato, yams | 3.9 | | 2.9 | | 7.4 | |
| Cassava | 3.4 | | 0.8 | | 2.4 | |
| | ___ | 13.9 | ___ | 9.7 | ___ | 6.9 |
| Legumes and oil seeds | | | | | | |
| Soybean | 4.2 | | 16.7 | | 39.8 | |
| Peanuts | 1.6 | | 4.8 | | 30.0 | |
| Peas | 1.3 | | 3.5 | | 26.9 | |
| Beans | 1.5 | | 5.4 | | 36.0 | |
| (Non-foods) | (1.6) | | (5.2) | | (35.4) | |
| | ___ | 10.2 | ___ | 35.6 | ___ | 34.9 |
| Other vegetables | | 2.8 | | 8.0 | | 28.6 |
| Fruit | | 2.5 | | 1.3 | | 5.2 |
| Animal products | | 10.2 | | 38.1 | | 37.4 |
| | | Total 152.8 | | Total 193.6 | | Mean 12.7 |

*Underlying concept*

An index of the balance of growth between root and shoot components of the plant integrated over a period of time. Effectively, a ratio between root and shoot mean relative growth rates (q.v.).

*Symbol and definition*

**K**; the allometric relationship is $R_W = bS_W^{\mathbf{K}}$, where $b$ is also a constant. The allometric coefficient, **K**, is the ratio between the mean RGRs of root and shoot, $\mathbf{R}_R$ and $\mathbf{R}_S$, i.e.

$$\mathbf{K} = \bar{\mathbf{R}}_R / \bar{\mathbf{R}}_S.$$

*Dimensions and units*

$T^{-1}T$, i.e. dimensionless. Equally balanced growth gives a **K** value of unity. In 'shooty' growth $\mathbf{K} < 1$ and in 'rooty' growth $\mathbf{K} > 1$. During the normal growth of herbaceous seedlings, the value of **K** often lies near 0.9.

*Formulae*

Because $R_W = bS_W^{\mathbf{K}}$,

$$\log R_W = \log b + \mathbf{K} \log S_W, \quad \text{or } \mathbf{K} = (\log R_W - \log b)/\log S_W.$$

*Methods of calculation*

Derived from a series of paired measurements of $R_W$ and $S_W$. The function $\log R_w = f(\log S_W)$ is fitted, usually as a linear regression of the form $\log R_W = \log b + \mathbf{K} \log S_W$. Strictly, a bivariate principal axis should be used because the $x$-variate ($S_W$) cannot be determined without error; however, the conventional regression is often employed.

*Relevant examples*

Root–shoot allometry in Italian ryegrass (Fig. 4.9) and in perennial ryegrass (Table 4.7).

For further details see CAUSTON & VENUS (pp. 173–218).

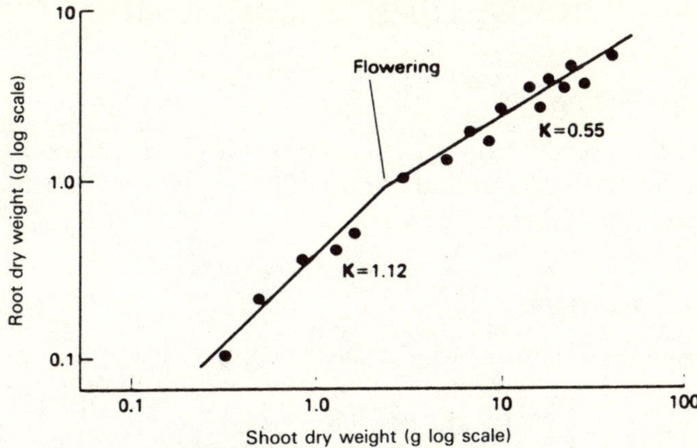


**Figure 4.9 Root–shoot allometry in Italian ryegrass** Flowering decreases the value of this allometric coefficient substantially. (From: A. Troughton (1956). *J. Br. Grassld. Soc.* 11, 56.)

**Table 4.7 Root–shoot allometry in perennial ryegrass** Values of **K**, the allometric coefficient, for the vegetative growth of perennial ryegrass under different conditions. 95% limits are given in parenthesis. (From: R. Hunt (1975). *Ann.Bot.* 39, 745.)

| | Full light | Shade |
|---|---|---|
| Full nutrients | 0.84 (0.04) | 0.69 (0.10) |
| Low nitrogen | 0.95 (0.04) | 0.73 (0.10) |

*Underlying concept*

An index of the balance of growth between any two components of the plant, integrated over a period of time. Effectively, a ratio between their mean relative growth rates (q.v.).

*Symbol and definition*

**K**; the allometric relationship is $Y = bZ^{\mathbf{K}}$, where $b$ is also a constant. The allometric coefficient, **K**, is the ratio between the mean RGRs of the plant components $Y$ and $Z$, $\mathbf{R}_Y$ and $\mathbf{R}_Z$, i.e.

$$\mathbf{K} = \bar{\mathbf{R}}_Y / \bar{\mathbf{R}}_Z.$$

where $Y$ and $Z$ may or may not be measured in the same units.

*Dimensions and units*

$T^{-1}T$, i.e. dimensionless. Equally balanced growth gives a **K** value of unity, whatever the units.

*Formulae*

Because $Y = bZ^{\mathbf{K}}$,

$$\log Y = \log b + \mathbf{K} \log Z, \quad \text{or } \mathbf{K} = (\log Y - \log b)/\log Z.$$

*Methods of calculation*

Derived from a series of paired measurements of $Y$ and $Z$. The function $\log Y = f(\log z)$ is fitted, usually as a linear regression of the form $\log Y = \log b + \mathbf{K} \log Z$. A bivariate principal axis is usually more appropriate than a conventional regression (see p. 50).

*Relevant examples*

Weight–area allometry in American willow-herb (Figure 4.10) and root–leaf allometry in maize (Fig. 4.11).

For further details see CAUSTON & VENUS (pp. 173–218); EVANS (pp. 268, 358).

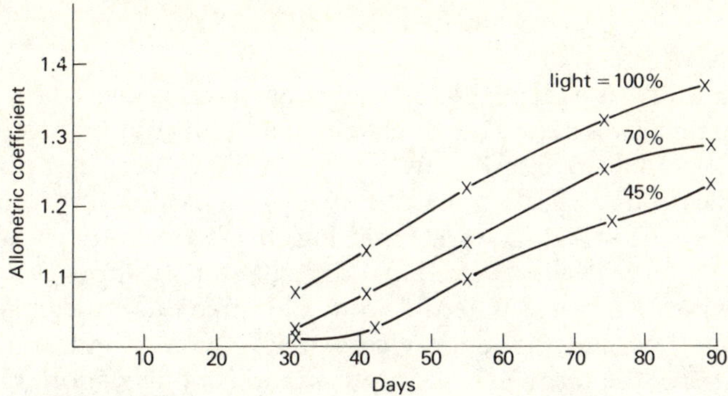


**Figure 4.10 Weight–area allometry in American willow-herb** The coefficient **K**, in the relation $W = bL_A{}^K$, drifts slowly upwards with time but is diminished by shading. (From: F. H. Whitehead & P. J. Myerscough (1962). *New Phytol*. 61, 314.)

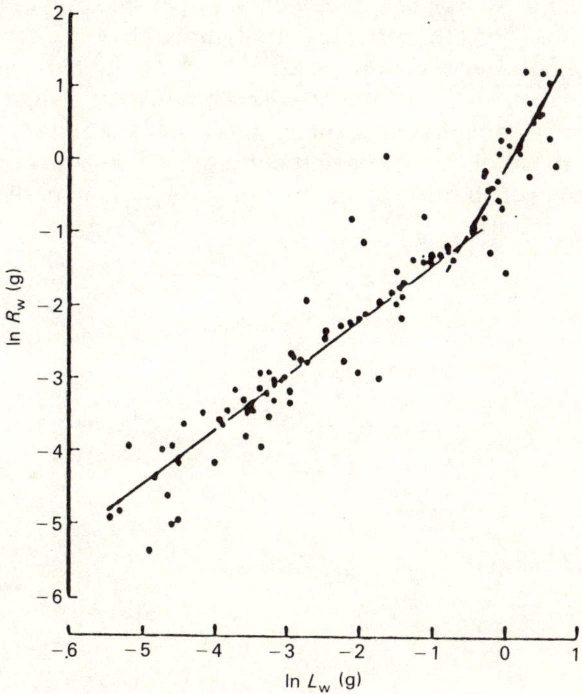


**Figure 4.11 Root–leaf allometry in maize** Towards the end of the growing period there is a cessation of leaf growth which causes a sharp increase in the value of **K** in the relation $R_W = bL_W{}^K$. (From: D. R. Causton & J. C. Venus (1981). *The Biometry of Plant Growth*. London: Edward Arnold.)

Mathematically, the Type III quantities are the simplest of all growth-analytical derivates. They alone are capable of instantaneous evaluation without recourse to fitted growth curves. However, if the functional approach is followed, $Z/Y$ should always be evaluated from separate functions such as $\log_e Z = f_Z(t)$ and $\log_e Y = f_Y(t)$, rather than directly from $Z/Y = f_{Z/Y}(t)$, which can have difficult statistical properties.

In the simplest cases, where $Z$ and $Y$ are like quantities (as with leaf weight ratio, efficiency of energy conversion, harvest index and harvestable protein), the term $Z/Y$ is simply an index of allocation. Opportunities to extend this variant of the concept are almost limitless.

Where $Z$ and $Y$ are unlike quantities (as with leaf area ratio, specific leaf area and leaf area index), the term $Z/Y$ is of interest as a 'snapshot' of the functional balance between two related or antagonistic components.

In the case of the allometric coefficients, it may be stretching a point to classify them as Type III derivates at all, since they are far from being simple ratios. However, as indices of the balance, not between sizes, but between growth rates (which may again be of like or unlike quantities), they possess the same intrinsic interest as other examples of the genre. There is also the added advantage that they often remain approximately constant across substantial intervals of time, which increases their value as comparative tools.

CHAPTER FIVE

# *Compounded growth rates*

*Underlying concept*

An index of the productive efficiency of plants calculated, not in relation to total dry weight (like RGR), but in relation to total leaf area. An index of leaf efficiency due to G. E. Briggs and co-workers. ULR is synonomous with net assimilation rate NAR, but the former term is older and better. ULR applies both to plants growing as spaced individuals and to plants growing in closed stands.

*Symbol and definition*

**E**; the rate of dry weight ($W$) production of a plant (or plant stand), expressed per unit of total leaf area $L_A$.

*Dimensions and units*

$\mathrm{M\,A^{-1}T^{-1}}$; typically $\mathrm{mg\,mm^{-2}\,day^{-1}}$ or $\mathrm{g\,m^{-2}\,day^{-1}}$.

*Formulae*

Instantaneously, $\mathbf{E} = (1/L_A)(\mathrm{d}W/\mathrm{d}t)$. The mean value over the interval $t_1$ to $t_2$ is approximately

$$\bar{\mathbf{E}} = (W_2 - W_1)/(t_2 - t_1) \times (\log_e L_{A2} - \log_e L_{A1})/(L_{A2} - L_{A1}).$$

*Method of calculation*

Instantaneously obtained from functions fitted to $\log_e W$ and to $\log_e L_A$ versus $t$; if $\log_e W = f_W(t)$ and $\log_e L_A = f_L(t)$, then $\mathbf{E} = (1/L_A)(\mathrm{d}W/\mathrm{d}t) = f_W'(t) \times \exp[f_W(t) - f_L(t)]$. Mean values are obtained from the formula for **E**, given above, using the separate estimates ($W_1$, $L_{A1}$) and ($W_2$, $L_{A2}$) made at times $t_1$ and $t_2$ respectively.

*Relevant examples*

Unit leaf rate in Kreusler's maize (Fig. 5.1) and seasonal changes in unit leaf rate (Fig. 5.2).

For relations to other quantities see pp. 90–95.

For further details see EVANS (pp. 200, 255, 333); CAUSTON & VENUS (p. 18); HUNT (p. 23).

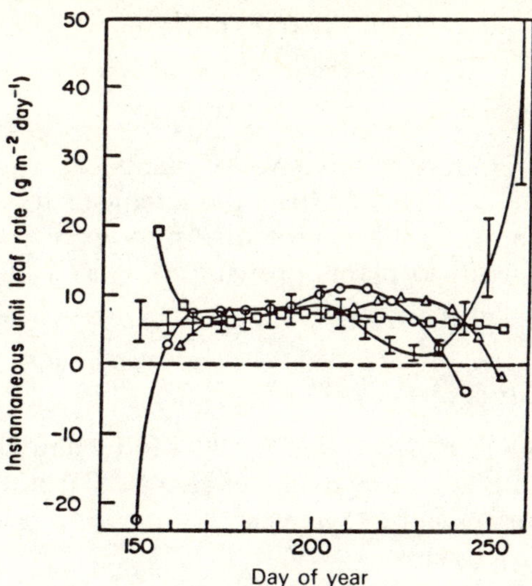


**Figure 5.1 Unit leaf rates in Kreusler's maize** Instantaneous values derived from fitted spline curves (see p. 105). The curve bearing 95% limits is for maize grown in 1875. Three other years' data are also shown: ( ○ ) 1876, ( □ ) 1877 and ( △ ) 1878 (see also Table 1.1 and Fig. 8.2). (From: R. Hunt & G. C. Evans (1980). *New Phytol*. 86, 155.)

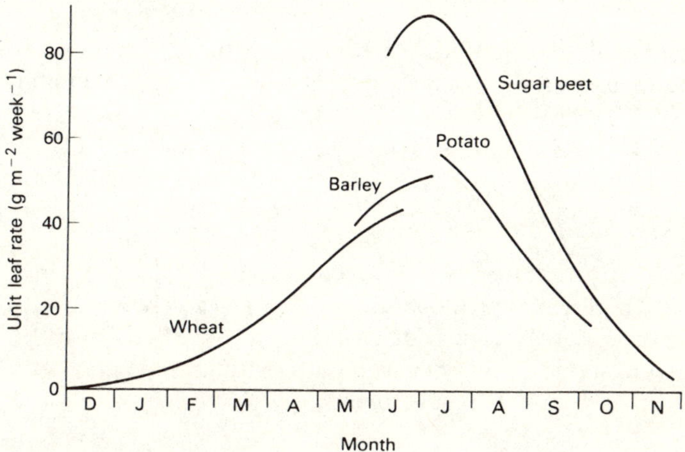


**Figure 5.2 Seasonal changes in unit leaf rate** Smoothed mean values for four crops grown at Rothamsted Experimental Station in the 1940s. Trends are highly dependent upon the seasonal environment, whatever the nature of the crop or the time of its establishment. (From: D. J. Watson (1947). *Ann. Bot*. 11, 41.)

*Underlying concept*

An index of the productive efficiency of plants calculated in relation to some measure of leaf size *other than* the total leaf, such as total leaf (or shoot) dry weight or total leaf protein. An index of leaf efficiency. Like ULR, applicable both to plants growing as spaced individuals and to plants growing in closed stands.

*Symbol and definition*

**B**, $\mathbf{E}_W$ and $\mathbf{E}_P$; the rate of dry weight ($W$) production of a plant or plant stand, expressed per unit of total leaf size, which in these three cases may be shoot dry weight ($S_W$), leaf dry weight ($L_W$) or leaf protein content ($L_P$) respectively.

*Dimensions and units*

$\mathrm{M\,M^{-1}T^{-1}}$; typically $\mathrm{mg\,mg^{-1}\,day^{-1}}$ or $\mathrm{g\,g^{-1}\,day^{-1}}$.

*Formulae*

Instantaneously $(1/L)(\mathrm{d}W/\mathrm{d}t)$, where $L$ is one of the measures of leaf size. The mean value over the interval $t_1$ to $t_2$ is approximately

$$\bar{\mathbf{E}} = (W_2 - W_1)/(t_2 - t_1) \times (\log_e L_2 - \log_e L_1)/(L_2 - L_1).$$

*Method of calculation*

Instantaneously obtained from functions fitted to $\log_e W$ and to $\log_e L$ versus $t$; if $\log_e W = f_W(t)$ and $\log_e L = f_L(t)$, then $(1/L)(\mathrm{d}W/\mathrm{d}t) = f_W'(t) \times \exp[f_W(t) - f_L(t)]$. Mean values are obtained from the formula for **E**, given above, using the separate estimates $(W_1, L_1)$ and $(W_2, L_2)$ made at times $t_1$ and $t_2$ respectively.

*Relevant examples*

Three kinds of unit leaf rate (Fig. 5.3) and unit shoot rate in perennial ryegrass (Fig. 5.4).
For further details see EVANS (p. 205); HUNT (p. 29).

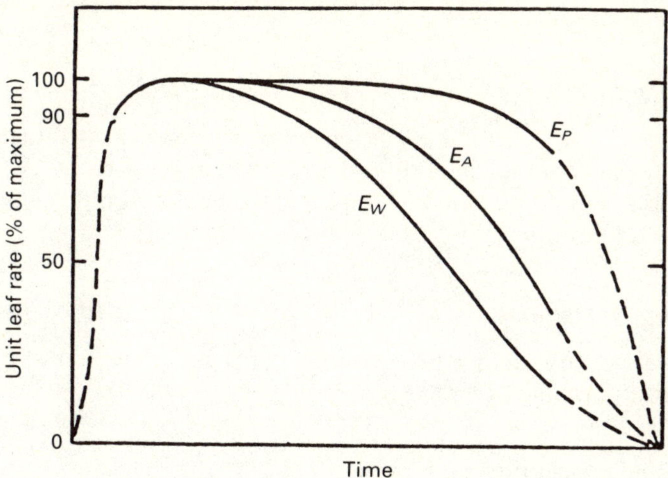


**Figure 5.3 Three kinds of unit leaf rate** When grown in a nitrogen-limited environment, plants may be expected to show an increasingly prolonged constancy of ULR according to the series shown here, calculated on the basis of leaf dry weight ($E_W$), leaf area ($E_A$) and leaf protein content ($E_P$). (From: R. F. Williams (1946). *Ann. Bot*. 10, 41.)

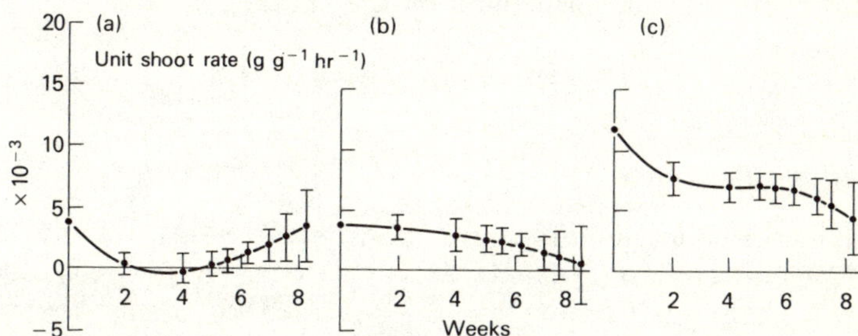


**Figure 5.4 Unit shoot rate in perennial ryegrass** Instantaneous values of the unit leaf rate on a total shoot weight basis decline gently with time under full light conditions (c), but are diminished, and made more constant in time, by 20% shade (b) and by 6% shade (a). (From: R. Hunt & J. A. Burnett (1973). *Ann. Bot*. 37, 519.)

*Underlying concept*

An index of the productive efficiency of land area in producing plant biomass. An index of agricultural productivity due to D. A. Watson. Applicable only to plants growing together in closed crops stands or in natural communities.

*Symbol and definition*

**C**; the rate of dry weight ($W$) production of the plant stand, expressed per unit of land area ($P$).

*Dimensions and units*

$\mathrm{M\,A^{-1}T^{-1}}$; typically $\mathrm{g\,m^{-2}\,day^{-1}}$ or $\mathrm{tonne\ ha^{-1}\,year^{-1}}$.

*Formulae*

Instantaneously $(1/P)(\mathrm{d}W/\mathrm{d}t)$. Because $P$ is not a plant growth variable, the usual formula for mean values of compounded growth rates does not apply. Mean values are given by the simpler formula

$$\bar{\mathbf{C}} = (1/P) \times (W_2 - W_1)/(t_2 - t_1).$$

*Method of calculation*

Instantaneously obtained from functions fitted to $\log_e (W/P)$ versus $t$; if $\log_e (W/P) = f(t)$ then $(1/P)(\mathrm{d}W/\mathrm{d}t) = f'(t) \times \exp[f(t)]$. Mean values are obtained from the formula for $\bar{\mathbf{C}}$, given above, using the separate estimates $W_1$ and $W_2$ obtained from equal ground areas $P$ at times $t_1$ and $t_2$ respectively. If ground areas are unequal at $t_1$ and $t_2$, then set $W_1 = W_1/P_1$ and $W_2 = W_2/P_2$ and use the formula $\bar{\mathbf{C}} = (W_2 - W_1)/(t_2 - t_1)$. In such a case, CGR is reduced to an absolute growth rate in size (q.v.).

*Relevant examples*

Crop growth rate in Kreusler's maize (Fig. 5.5) and crop growth rates of temperate crops (Table 5.1).

For relations to other quantities see pp. 92–97.

For further details see HUNT (p. 35).

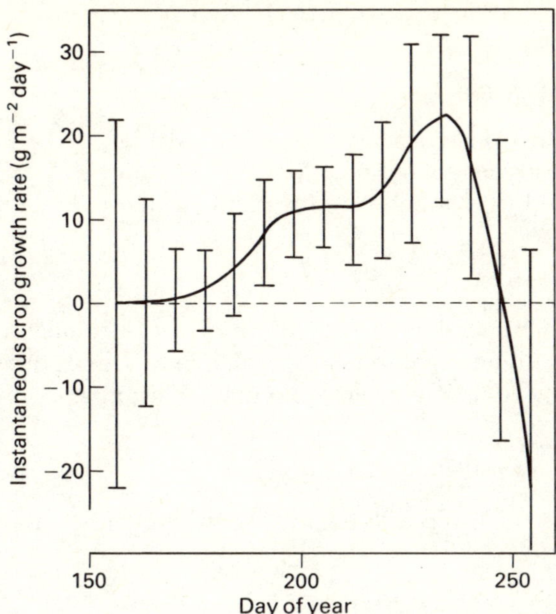


**Figure 5.5 Crop growth rate in Kreusler's maize** Instantaneous values derived from splined curves (see p. 104) fitted to data from the 1877 crop. Because the density of individuals within the stand was almost constant thoughout, this curve closely resembles the progression of mean absolute growth rate per individual (shown for the 1878 data in Fig. 2.1). (R. Hunt & G. C. Evans, unpubl.)

**Table 5.1 Short-term crop growth rates** Rates of dry weight yield of temperate crops and short-term photosynthetic efficiencies. (From: COOMBS & HALL, p. 160.)

| Crop | Location | Crop growth rate (g m$^{-2}$ day$^{-1}$) | Photosynthetic efficiency (% of total radiation) |
|---|---|---|---|
| Tall fescue | UK | 43 | 3.5 |
| Ryegrass | UK | 28 | 2.5 |
| Cocksfoot | UK | 40 | 3.3 |
| Sugar beet | UK | 31 | 4.3 |
| Kale | UK | 21 | 2.2 |
| Barley | UK | 23 | 1.8 |
| Maize | UK | 24 | 3.4 |
| Wheat | Netherlands | 18 | 1.7 |
| Peas | Netherlands | 20 | 1.9 |
| Red clover | New Zealand | 23 | 1.9 |
| Maize | New Zealand | 29 | 2.7 |
| Maize | USA, Kentucky | 40 | 3.4 |

*Underlying concept*

An index of the uptake efficiency of roots, calculated in relation to some measure of root size. An index of root activity due to P. J. Welbank. Valid both for spaced individuals and for plants growing in closed stands.

*Symbol and definition*

**A**; the rate of mineral nutrient ($M$) uptake of a plant or plant stand expressed per unit of total root size, which may be root dry weight ($R_W$), or alternatively root length, area, volume or number.

*Dimensions and units*

$M M^{-1} T^{-1}$ (for $R_W$); typically ug mg$^{-1}$ day$^{-1}$ or mg g$^{-1}$ day$^{-1}$.

*Formulae*

Instantaneously, $\mathbf{A} = (1/R)(\mathrm{d}M/\mathrm{d}t)$, where $R$ is one of the measures of root size. The mean value over the internal $t_1$ to $t_2$ is approximately

$$\bar{\mathbf{A}} = (M_2 - M_1)/(t_2 - t_1) \times (\log_e R_2 - \log_e R_1)/(R_2 - R_1),$$

where $M$ may be the combined contents of more than one mineral nutrient element.

*Method of calculation*

Instantaneously obtained from functions fitted to $\log_e M$ and to $\log_e R$ versus $t$; if $\log_e M = f_M(t)$ and $\log_e R = f_R(t)$, then $\mathbf{A} = (1/R)(\mathrm{d}M/\mathrm{d}t) = f_M'(t) \times \exp[f_M(t) - f_R(t)]$. Mean values are obtained from the formula for $\mathbf{A}$, given above, using the separate estimates ($M_1$, $R_1$) and ($M_2$, $R_2$) made at times $t_1$ and $t_2$ respectively.

*Relevant examples*

Specific absorption rate in sugar beet (Fig. 5.6) and in cultivated cranberry (Fig. 5.7).

For further details see EVANS (p. 228); HUNT (p. 30).

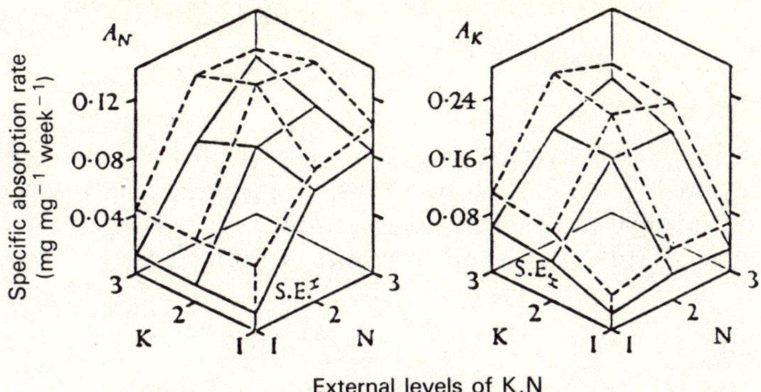


**Figure 5.6 Specific absorption rate in sugar beet** Calculated for nitrogen ($A_N$) and potassium ($A_K$) over a two-week period. Broken lines: sugar beet grown alone; solid lines, beet grown in competition with couch grass. (From: P. J. Welbank (1964). *Ann. Bot*. 28, 1.)

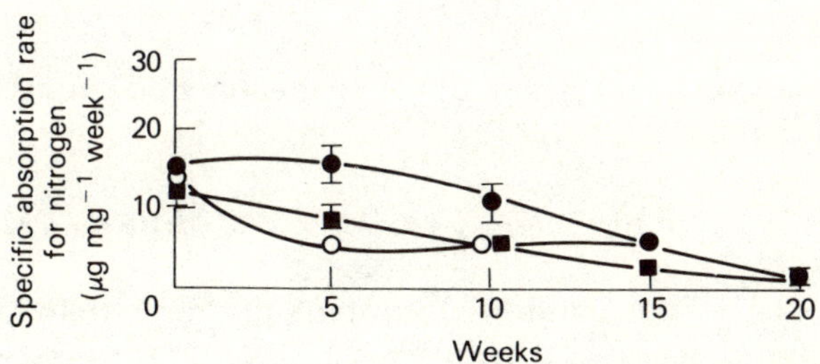


**Figure 5.7 Specific absorption rate in cultivated cranberry** Rates of nitrogen uptake for mycorrhizal plants grown on a poor native soil ( ■ ), mycorrhizal plants grown on the irradiated soil (•), and non-mycorrhizal plants grown on the irradiated soil ( o ). Mycorrhizal infection increases nitrogen uptake, but soil irradiation does not. (From: Stribley *et al*. (1975). *New Phytol*. 75, 119.)

*Underlying concept*

An index of the productive efficiency of mineral nutrients within the plant; due to J. Keay and co-workers. A measure of the dry weight returns on mineral nutrient uptake, a nutrient productivity. Applicable both to spaced individuals and to plants growing in closed stands.

*Symbol and definition*

**U**; the rate of dry weight ($W$) production of a plant or plant stand expressed per unit of mineral nutrient content ($M$).

*Dimensions and units*

$\mathrm{M\,M^{-1}\,T^{-1}}$; typically $\mathrm{mg\,mg^{-1}\,day^{-1}}$ or $\mathrm{g\,g^{-1}\,day^{-1}}$.

*Formulae*

Instantaneously, $\mathbf{U} = (1/M)(\mathrm{d}W/\mathrm{d}t)$. The mean value over the interval $t_1$ to $t_2$ is approximately

$$\bar{\mathbf{U}} = (W_2 - W_1)/(t_2 - t_1) \times (\log_e M_2 - \log_e M_1)/(M_2 - M_1),$$

where $M$ may be the combined content of more than one mineral nutrient element.

*Method of calculation*

Instantaneously obtained from functions fitted to $\log_e W$ and to $\log_e M$ versus $t$; if $\log_e W = f_W(t)$ and $\log_e M = f_M(t)$, then $\mathbf{U} = (1/M)(\mathrm{d}W/\mathrm{d}t) = f_W'(t) \times \exp[f_W(t) - f_M(t)]$. Mean values are obtained from the formula for $\bar{\mathbf{U}}$, given above, using the separate estimates $(W_1, M_1)$ and $(W_2, M_2)$ made at times $t_1$ and $t_2$ respectively.

*Relevant examples*

Specific utilization rate in Australian pasture species (Table 5.2) and in coniferous saplings (Table 5.3).

For further details see HUNT (p. 30).

**Table 5.2 Specific utilization rate in Australian pasture species** Rate of production of dry matter per unit of phosphorus, g (mg atom P)$^{-1}$ day$^{-1}$. (From: J. Keay *et. al.* (1979). *Aust. J. agric. Res.* 21, 33.)

| Species | Harvest interval (days) | | |
|---|---|---|---|
| | 1–29 | 29–57 | 57–92 |
| Cupped clover | 4.3 | 0.85 | 1.0 |
| Subterranean clover | 4.2 | 0.92 | 1.2 |
| Rose clover | 4.0 | 0.94 | 0.93 |
| Silver grass | 3.5 | 1.4 | 1.3 |
| Wimmera ryegrass | 3.3 | 2.0 | 0.99 |
| Cape-weed | 3.1 | 2.0 | 1.2 |
| Erodium | 2.9 | 1.1 | 0.75 |
| Lupin | 1.6 | 0.63 | 0.94 |

**Table 5.3 Specific utilization rate in coniferous saplings** Maximum relative growth rate and SUR, or 'nitrogen productivity'. (From: T. Ingestad & M. Kahr (1985). *Physiologia Pl.* 65, 109.)

| Species | Maximum RGR (% day$^{-1}$) | Nitrogen productivity (g g$^{-1}$ hour$^{-1}$) |
|---|---|---|
| Lodgepole pine | 7.4 | 0.16 |
| Scots pine | | |
| Southern provenance | 7.4 | 0.14 |
| Northern provenance | 7.3 | 0.14 |
| Norway spruce | | |
| Southern provenance | 7.1 | 0.15 |
| Northern provenance | 6.1 | 0.14 |

*Underlying concept*

Within the plant cell, the pathway DNA → RNA → protein nitrogen → cell wall material is accessible to the calculation of indices of productive efficiency. Each index demonstrates the efficiency of one component in the series in creating the next.

*Symbol and definition*

$\mathbf{G}_{i,i-1}$; the rate of dry weight production of the *i*th component of the plant cell series, expressed per unit of component $i-1$; e.g. $\mathbf{G}_{\text{RNA,DNA}}$, the rate of production of RNA per unit of DNA.

*Dimensions and units*

$\text{M}\,\text{M}^{-1}\text{T}^{-1}$; typically $\text{mg}\,\text{mg}^{-1}\,\text{day}^{-1}$.

*Formulae*

Instantaneously $(1/w_{i-1})(\text{d}w_i/\text{d}t)$, where $w_1$ and $w_{i-1}$ are component weights at stage $i$ and $i-1$ respectively. The mean value over the interval $t_{(1)}$ to $t_{(2)}$ is approximately

$$\overline{\mathbf{G}}_{i,\,i-1} = (w_{i(2)} - w_{i(1)})/(t_{(2)} - t_{(1)}) \\ \times (\log_e w_{i-1(2)} - \log_e w_{i-1(1)})/(w_{i-1(2)} - w_{i-1(1)}).$$

*Method of calculation*

Instantaneously obtained from functions fitted to $\log_e w_i$ and to $\log_e w_{i-1}$ versus $t$; if $\log_e w_i = f_i(t)$ and $\log_e w_{i-1} = f_{i-1}(t)$, then $\mathbf{G}_{i,\,i-1} = (1/w_{i-1})(\text{d}w_i/\text{d}t) = f_i'(t) \times \exp[f_i(t) - f_{i-1}(t)]$. Mean values are obtained from the formula for $\overline{\mathbf{G}}$, given above, using the separate estimates $(w_{i(1)}, w_{i-1(1)})$ and $(w_{i(2)}, w_{i-1(2)})$ made at times $t_{(1)}$ and $t_{(2)}$ respectively.

*Relevant examples*

Sub-cellular efficiencies in clover (Table 5.4).

For further details see HUNT (p. 31) and the reference cited in the example.

**Table 5.4 Sub-cellular efficiencies in clover** Rates of production, G. of one component of the leaf per unit of another component in subterranean clover. (From: R. F. Williams (1975). *The Shoot Apex and Leaf Growth*. Cambridge: Cambridge University Press.)

| Interval (days) | $G_{RNA,DNA}$ ($day^{-1}$) | $G_{PN,RNA}$ ($day^{-1}$) | $G_{CW,PN}$ ($day^{-1}$) | $G_{CW,CS}$ ($ng\ mm^{-2}\ day^{-1}$) |
|---|---|---|---|---|
| 10–11 | 6.5 | 5.8 | 0.56 | 117 |
| 11–12 | 7.4 | 5.7 | 0.95 | 182 |
| 12–13 | 6.0 | 5.3 | 1.29 | 224 |
| 13–14 | 7.5 | 4.1 | 1.64 | 234 |
| 14–15 | 6.4 | 5.6 | 1.36 | 177 |
| 15–16 | 6.9 | 4.7 | 1.13 | 157 |
| 16–17 | 3.6 | 4.4 | 0.83 | 127 |
| 17–19 | 4.6 | 4.4 | 0.58 | 111 |
| 19–22 | — | 1.8 | 0.23 | 52 |
| 22–25 | — | 1.1 | 0.18 | 43 |
| 25–29 | — | — | 0.32 | 70 |
| 29–36 | — | — | 0.07 | — |

PN, protein nitrogen; CW, cell wall material; CS, cell surface, including leaf hairs

*Underlying concept*

In deciduous trees part of the total biomass is summer-temporary. RGRs including all of the current biomass may therefore underestimate the efficiency of the truly perennating part of the tree's material. CPR is an index of the productive efficiency of the perennating biomass, due to P. J. Dudney. It is an RGR-substitute, specially devised for perennial plants.

*Symbol and definition*

$\mathbf{\Pi}$; the rate of dry weight ($W$) production of a tree or tree stand expressed per unit of perennating structure ($W_P$).

*Dimensions and units*

$M\,M^{-1}T^{-1}$; typically $g\,g^{-1}\,day^{-1}$ or $kg\,kg^{-1}\,week^{-1}$.

*Formulae*

Instantaneously, $\mathbf{\Pi} = (1/W_P)(dW/dt)$. The mean value over the interval $t_1$ to $t_2$ is approximately

$$\bar{\mathbf{\Pi}} = (W_2 - W_1)/(t_2 - t_1) \times (\log_e W_{P2} - \log_e W_{P1})/(W_{P2} - W_{P1}).$$

*Method of calculation*

Instantaneously obtained from functions fitted to $\log_e W$ and to $\log_e W_P$ versus $t$; if $\log_e W = f_W(t)$ and $\log_e W_P = f_P(t)$, then $\mathbf{\Pi} = (1/W_P)(dW/dt) = f_W'(t) \times \exp[f_W(t) - f_P(t)]$. Mean values are obtained from the formula for $\bar{\mathbf{\Pi}}$, given above, using the separate estimates ($W_1$, $W_{P1}$) and ($W_2$, $W_{P2}$) made at times $t_1$ and $t_2$ respectively.

*Relevant examples*

Unit production rate in apple trees (Fig. 5.8).

For further details see HUNT (p. 31).

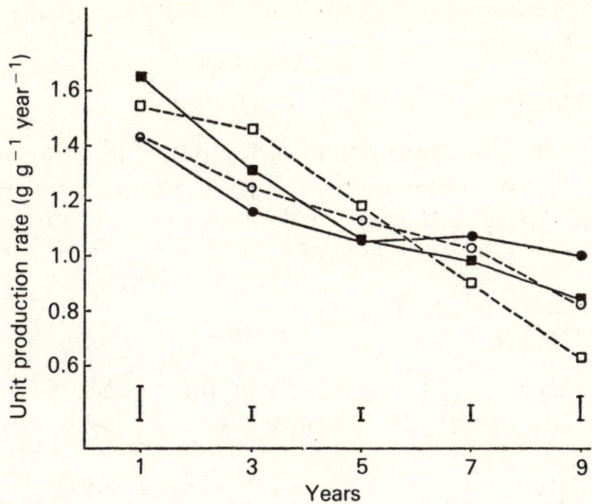


**Figure 5.8 Unit production rate in apple trees** Round symbols, Cox variety M26; square symbols, Cox variety MM104. Open symbols, management by minimal-pruning; closed symbols, management by spur-pruning. Efficiency of perennating structures declines with time, but is enhanced by the more radical spur-pruning. (From: P. J. Dudney (1974). *Ann. Bot.* 38, 647.)

### *Underlying concept*

An index of the current commitment of the whole plant to the production of one of its components, such as roots, stem or individual leaves. Applicable both to spaced individuals and to plants growing in closed stands.

### *Symbol and definition*

$\mathbf{J}_i$; the rate of dry weight production of the *i*th plant component ($w_i$) expressed per unit of total dry weight ($W$).

### *Dimensions and units*

$\mathrm{M\,M^{-1}T^{-1}}$; typically $\mathrm{mg\,mg^{-1}\,day^{-1}}$ or $\mathrm{g\,g^{-1}\,day^{-1}}$.

### *Formulae*

Instantaneously, $\mathbf{J}_i = (1/W)(\mathrm{d}w_i/\mathrm{d}t)$. The mean value over the interval $t_1$ to $t_2$ is approximately

$$\bar{\mathbf{J}}_i = (w_{i2} - w_{i1})/(t_2 - t_1) \times (\log_e W_2 - \log_e W_1)/(W_2 - W_1).$$

### *Method of calculation*

Instantaneously obtained from functions fitted to $\log_e w_i$ and to $\log_e W$ versus $t$; if $\log_e w_i = f_w(t)$ and $\log_e W = f_W(t)$, then $\mathbf{J} = (1/W)(\mathrm{d}w_i/\mathrm{d}t) = f_w{}'(t) \times \exp[f_w(t) - f_W(t)]$. Mean values are obtained from the formula for $\bar{\mathbf{J}}$, given above, using the separate estimates ($w_{i1}$, $W_1$) and ($W_{i2}$, $W_2$) made at times $t_1$ and $t_2$ respectively.

### *Relevant examples*

Component production rates in giant ragweed (Fig. 5.9).

For relations to other quantities see pp. 88–89.

For further details see HUNT (p. 32).

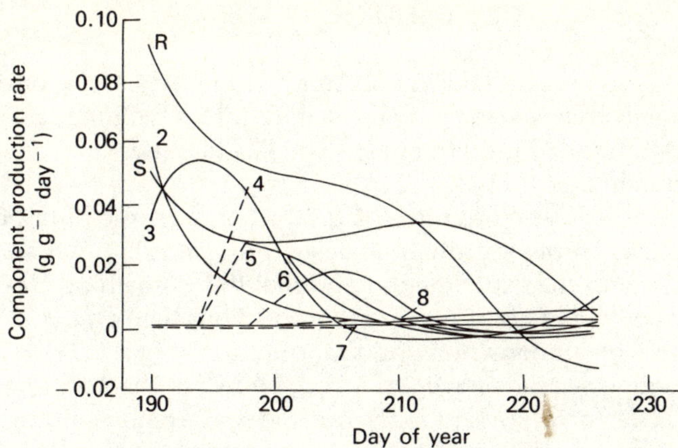


**Figure 5.9 Component production rates in giant ragweed** Rates of production for roots (R), stem material (S) and successive leaves (at nodes 1 to 8), illustrating the changing commitment of the plant to different forms of vegetative growth (see also Fig. 8.1). (From: R. Hunt & F. A. Bazzaz (1980). *New Phytol*. 84, 113.)

If $Y$ and $Z$ are again general symbols representing any of the primary data in plant growth analysis (the various weights, areas, volumes or numbers), then the compounded growth rates, Type IV quantities, may be given the general notation $(1/Z)(\mathrm{d}Y/\mathrm{d}t)$.

In plain words, these are rates of production of something per unit of something else. In plant growth analysis, provided that the 'something', $Y$, is of interest to the experimenter and that the 'something else', $Z$, may reasonably be held responsible for its production, then $(1/Z)(\mathrm{d}Y/\mathrm{d}t)$ is an analytical tool of fundamental importance.

The preceding sections have described the use of this tool in a selection of guises. The form $(1/Z)(\mathrm{d}Y/\mathrm{d}t)$ can aid our understanding of plant processes from the molecular level ($\mathbf{G}_{i,\,i-1}$), through organs ($\mathbf{J}$), and whole herbaceous plants ($\mathbf{A}$, $\mathbf{B}$, $\mathbf{E}$, $\mathbf{U}$) to perennial woody crops ($\mathbf{\Pi}$).

The list given is not exhaustive; nor have all of the possibilities in this direction been followed up. The reader may be stimulated to invent a few more and evaluate them either instantaneously or as harvest-interval means using analogues of the formula on p. 56. Provided that the above caveat is borne in mind, there is no reason why the applications of $(1/Z)(\mathrm{d}Y/\mathrm{d}t)$ should not flourish still further.

# CHAPTER SIX

# *Integral durations*

## *Underlying concept*

When leaf area index (LAI), or total leaf area per crop, is plotted against time, the area beneath the curve shows the 'whole opportunity for assimilation' which the crop possesses (D. J. Watson). LAD takes into account not only how much leaf area develops, but also how long it lasts. LAD is thus expressed per crop or per season.

## *Symbol and definition*

**D**; the integral of the area beneath (i) the curve of LAI versus time; or (ii) the curve of total leaf area per crop, $L_A$, versus time.

## *Dimensions and units*

(i) $[\mathrm{A\,A^{-1}}]\,\mathrm{T}$, (ii) $\mathrm{A\,T}$; typically (i) weeks, or (ii) $m^2$ weeks.

## *Formulae*

An integral, not an instantaneous, quantity; (i) $\mathbf{D} = {}_{t1}\int^{t2} \mathbf{L}\mathrm{d}t$, or (ii) $\mathbf{D} = {}_{t1}\int^{t2} L_A \,\mathrm{d}t$. A value over the interval $t_1$ to $t_2$ is given by

$$\text{(i)}\quad \mathbf{D} = (L_1 + L_2)(t_2 - t_1)/2, \quad \text{or} \quad \text{(ii)}\quad \mathbf{D} = (L_{A1} + L_{A2})(t_2 - t_1)/2.$$

## *Methods of calculation*

Integration of (i) the function $\mathbf{L} = f_{LAI}(t)$, or (ii) the function $L_A = f_{LA}(t)$, between the limits $t_1$ and $t_2$; this may be done either numerically or graphically. The value over the interval $t_1$ to $t_2$ may be calculated directly from the above formula if ($\mathbf{L}_1$, $\mathbf{L}_2$) or ($L_{A1}$, $L_{A1}$) are available.

## *Relevant examples*

Leaf area duration in wheat (Fig. 6.1) and in *Typha* and *Phragmites* (Table 6.1).

For relations to other quantities see pp. 94–95.

For further details see CAUSTON (p. 112); EVANS (p. 223); HUNT (p. 37).

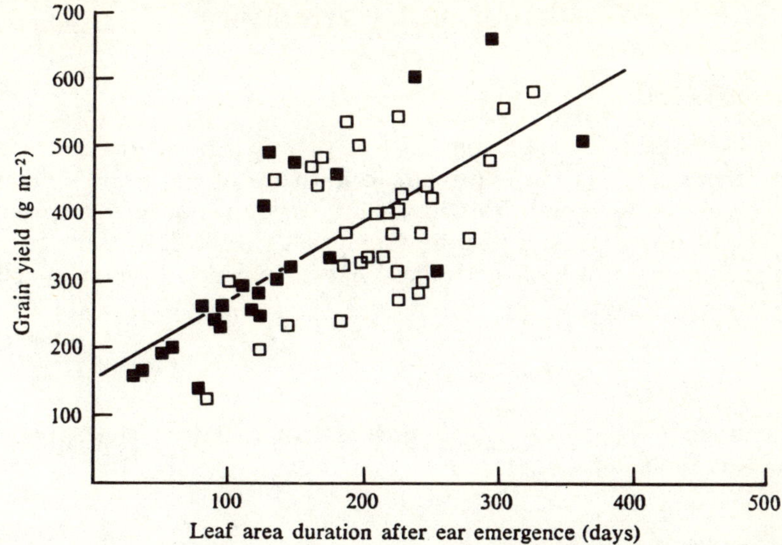


**Figure 6.1 Leaf area duration in wheat** The relation between grain yield and leaf area duration after ear emergence. Open symbols, UK crops; closed symbols, sub-tropical crops. (From: L. T. Evans *et al*. (1975). In: L. T. Evans (ed.), *Crop Physiology: Some Case Histories*. Cambridge: Cambridge University Press.)

**Table 6.1 Leaf area duration in *Typha* and *Phragmites* reed beds** Values are calculated for a whole growing season for different strata in the reed canopy. (From: J. Květ *et al*. (1969). *Hidrobiologia* 10, 63.)

| Canopy strata (m height) | *Typha* LAD (days) | *Phragmites* LAD (days) |
|---|---|---|
| 3.6–4.0 | — | 0.39 |
| 3.2–3.6 | — | 14.50 |
| 2.8–3.2 | — | 66.30 |
| 2.4–2.8 | 0.28 | 142.99 |
| 2.0–2.4 | 21.76 | 188.22 |
| 1.6–2.0 | 64.92 | 147.31 |
| 1.2–1.6 | 107.94 | 75.20 |
| 0.8–1.2 | 117.00 | 30.70 |
| 0.4–0.8 | 84.75 | 5.91 |
| 0–0.4 | — | 0.11 |
| Whole canopy | 396.65 | 671.63 |

*Underlying concept*

When the total dry weight per crop is plotted against time, the area beneath the curve is 'an approximate measure of the stand's vitality' (J. Kvet and co-workers). BMD takes into account not only how much dry weight develops, but also how long it lasts. BMD is thus the dry weight equivalent of LAD and, like LAD, is expressed per crop or per season.

*Symbol and definition*

**Z**; the integral of the area beneath the curve of total dry weight per crop, *W*, versus time.

*Dimensions and units*

M T; typically g weeks.

*Formulae*

An integral, not an instantaneous, quantity; $\mathbf{D} = {}_{t1}\int^{t2} W dt$. A value over the interval $t_1$ to $t_2$ is given by

$$\mathbf{D} = (W_1 + W_2)(t_2 - t_1)/2.$$

*Methods of calculation*

Integration of the function $W = f(t)$ between the limits $t_1$ and $t_2$; this may be done either numerically or graphically. The value over the interval $t_1$ to $t_2$ may be calculated directly from the above formula if ($W_1$, $W_2$) are available.

*Relevant examples*

The concept of biomass duration (Fig. 6.2) and biomass duration in a woodland ground flora (Table 6.2).

For relations to other quantities see p. 94.

For further details see CAUSTON (p. 112); HUNT (p. 39).

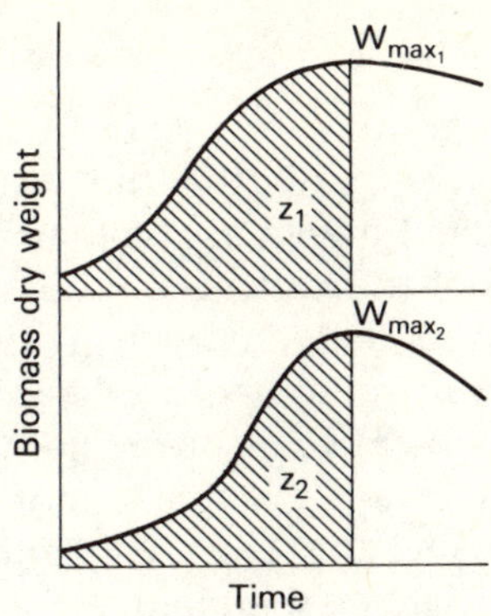


**Figure 6.2 The concept of biomass duration** Biomass in two different stands may attain the same maximum value ($W_{max1} = W_{max2}$) by growth along two different curves. This leads to two different biomass durations, the areas $Z_1$, $Z_2$. (From: J. Květ & J. P. Ondok (1971). *Photosynthetica* 5, 417.)

**Table 6.2 Biomass duration in a woodland ground flora** $W_{max}$ is the seasonal maximum shoot biomass, $t$ is the length of the period from shoot emergence to $W_{max}$, **Z** is shoot BMD (all $m^{-2}$). The species are from a mixed deciduous woodland community in Central Europe. (After: J. Květ & J. P. Ondok (1971). *Photosynthetica* 5, 417.)

| Species | $W_{max}$ (g) | $t$ (d) | Z (g d) |
|---|---|---|---|
| *Anemone nemorosa* | 20.0 | 33 | 393 |
| *Isopyrum thalictroides* | 2.8 | 49 | 53 |
| *Primula elatior* | 5.6 | 134 | 490 |
| *Lathyrus vernus* | 1.3 | 155 | 138 |
| *Aegopodium podagraria* | 12.0 | 162 | 1620 |
| *Pulmonaria officinalis* | 18.0 | 162 | 2270 |
| *Lamium galeobdolon* | 2.8 | 162 | 200 |
| *Ranunculus cassubicus* | 1.0 | 162 | 140 |
| *Fragaria elatior* | 0.4 | 162 | 35 |
| *Hepatica nobilis* | 2.3 | 200 | 300 |
| *Asarum europaeum* | 2.3 | 200 | 300 |
| Other spp. | 0.4 | 162 | 40 |
| All spp. | 66.6 | 162* | 5679 |

*The effective growing season of the community.

## CONCLUDING REMARKS

The Type V derived quantities, or integral durations (pp. 15–16), are mathematically unlike any other in that they exist only as areas beneath functions between definite time limits. Biologically, they summarize the whole presence in time of the variables in question, whether these are leaf area (index) or biomass. Fitted curves provide an excellent basis for the derivation of integral durations, as the computation of areas by graphical methods, of which Simpson's rule (CAUSTON, p. 126) is a sophisticated variant, is always to some extent approximate.

The possibilities for using the Type V quantity in conjunction with other plant variates are relatively limited. Perhaps the most promising opportunities lie in cases where it is necessary to estimate just how much of a scarce environmental resource, such as mineral nutrient(s) or water, is tied up in plant material, and for how long. For example, a 'nutrient duration' comprising the integral of the curve of cumulative plant nutrient content might provide useful clues as to the competitive mechanisms operating in closed natural communities or in crop/weed systems.

CHAPTER SEVEN

# *Other independent variables*

Rates of change are, of course, rates seen in relation to time. Sometimes, progressions of rates or ratios may be plotted against quantities other than time.

The effects of any experimental treatments may accelerate or decelerate the natural time-based drift in the quantity under examination. Hence, plants of equal age but of unequal experimental history are not necessarily in the same morphogenetic condition as one another. To overcome this disadvantage (minimizing the effects of ontogenetic drift) it can be useful to plot the derived quantities against other indices of development, such as total dry weight or number of leaves, instead of against time.

In practice this may be done by plotting values of derived quantities, like RGR and ULR, in order of, say, increasing $W$. Naturally, this is easier with the instantaneous values than with the mean ones. Indeed, if dry weight does not figure in the derived quantity itself (as with SAR, which is $(1/R_W)(dM/dT)$), $W$ may actually be inserted into the analyses in place of $T$ (giving $(1/R_W)(dM/dW)$).

However, these arguments are as broad as they are long: what guarantee is there that plants of equal dry weight but in different treatments have reached that dry weight by similar routes and are currently in the same morphogenetic condition?

There is also another danger here. When quantities derived from $W$ are plotted against $W$, or against other quantities derived from it, misleading relationships can be obtained. If, for example, RGR is roughly constant, plotting ULR against LAR (that is, against RGR/ULR, see p. 90) can produce a straightish line for mathematical reasons alone. This representation can obscure a quite genuine dependence of ULR on LAR when increases in LAR lead to self-shading, thus causing a fall in ULR (see also Fig. 8.4, p. 95).

One developmental time scale is the plastochron index. A plastochron is (normally) the period of time between the initiation of one leaf and the next within a shoot apex. Plant age on the basis of the plastochron is expressed as the number of such intervals that have elapsed. Some have found disadvantages attached to the use of a time-based abscissa when investigating variates such as fresh weight, chlorophyll content and respiratory behaviour in successive individual leaves. For these, convincing progressions can be assembled on a plastochron 'time' scale.

Various attempts have also been made to construct a developmental scale involving other organs, in addition to leaves. A decimal scale of principal and secondary stages of growth, potentially 100 points long, has been devised for cereals. Stylized drawings are available of selected stages in the growth of wheat, barley and oats, with expanded descrip-

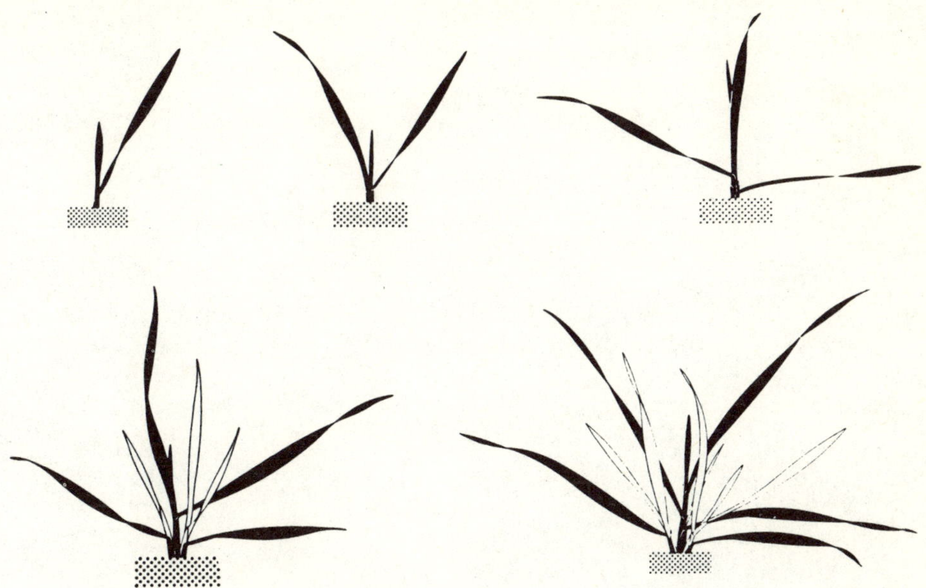

**Figure 7.1 Growth stages in wheat** A 100-point decimal scale exists, running from seed to seed through the entire life cycle of a cereal. The five stages illustrated here are differentiated by number of leaves unfolded by the initial tiller, i.e. 1 to 5. These stages represent points 11 to 15 on the decimal scale. (From: D. R. Tottman *et al*. (1979). *Ann. appl. Biol.* 93, 221.)

tions of some of the stages (Fig. 7.1). Numbers of root nodules in legumes have also been put forward for use as a developmental scale.

In addition to methods which recognize the natural developmental stages of the plant, some have been devised to take into account the accumulated experience of the plant in respect of selected environmental variables. J. A. Nelder devised a scheme with the general aim of 'replacing chronological time in the growth equations of vegetative crops by a time scale based on some suitable combination of meteorological factors.' Others have considered growth in relation to the level of available photosynthetically active radiation and tested the utility of logistic models fitted on a variety of heat unit time scales (calculated as temperature integrals above a range of base temperatures derived from daily maxima and minima).

A substantial field of activity also exists in which other independent variables or variates are used not instead of, but in addition to, time. An example of this approach appears at the end of the next chapter.

CHAPTER EIGHT

# *Interrelations*

Although the quantities involved in plant growth analysis each have an independent meaning, as described in the preceding sections, their strength as analytical tools owes much to their interrelations with one another. Various types of interrelations can occur, but they all share the same general properties which arise from their special mathematical status.

### *Identities and conditional equations*

Growth analysis uses equations of a type known as 'identities'. These are fundamentally distinct from the 'conditional equations' which appear in some other mathematical models of growth. Both types of equation are algebraic statements of equality between two expressions, but there is a mathematical distinction between them.

An **identity** in a statement of equality which holds true for all meaningful values of the variables; for example, $x^2 - y^2 = (x + y)(x - y)$, and $a/b = az/bz$. (To be 'meaningful', $b$ and $z$ must be non-zero.)

A **conditional equation** is a statement of equality which holds true for some, but not all, meaningful values of the variables; for example, $2x + y = 8$, and $xy + 2z = 10$.

An identity can be proved logically, for it is algebraically self-evident, or tautological. This is not so for a conditional equation, which has a hypothetical or empirical content; it can be tested by observation or experiment but it cannot be proven.

### *Models of plant growth*

All the growth-analytical interrelations described in this chapter are identities of the algebraically self-evident type. In contrast, other mathematical models of growth use conditional equations. Simple examples of these are (a) the equation assuming exponential growth, $W = W_0e^{rt}$, where $W_0$ and $W$ are the dry weights of a plant before and after growth at the exponential rate $r$ during the time interval $t$; and (b) the equation for the hypothesis that the crop growth rate is linearly dependent upon the light energy received, $N\mathrm{d}W/\mathrm{d}t = kI_0$, where $N$ is number of plants per unit ground area, $I_0$ is incident light flux density and $k$ is a constant.

*Consequences*

Although identities and conditional equations have much in common, there are three basic consequences for plant growth studies which arise from the differences which lie between them.

First, an identity is tautological in the strict sense of the word; it states a logical truth. Hence, there is no point in attempting to validate such an equation. A conditional equation used in a model of growth is, however, an hypothesis. It expresses a supposed relation between certain quantities in mathematical terms. It may be universally true, true only approximately, or not true at all. The testing, evaluation or validation of such an equation is essential.

Second, conditional equations usually include constants which have the same fixed values in all applications of the model (changes in them may present problems). There are also variable parameters present which are of particular significance only to the current application of the model. By contrast, the identities of plant growth analysis include no constants or parameters and are based wholly upon the primary variates. The interest, however, lies less in the values of these primary variates than in those of the derived quantities which feature in the equation, such as RGR, ULR, etc. Differences between these with respect to genotype, environment, age or treatment are of prime importance.

Third, conditional equations in growth models can be used to quantify one known if all other quantities in the equation are known. Often, such equations are used to calculate a dependent variable. This procedure can be relied upon *only* if the equation is known to be valid. On the other hand, growth analytical identities cannot be used to calculate values of primary variates such as $W$, or $L_A$, since the equations are true for *all* values of these variates. However, such an equation can be used as a defining relation to calculate an unknown derived quantity if all the other components are known. This offers, for example, a unique method of deriving instantaneous values of quantities such as ULR, for which only mean values over a period can otherwise be measured.

Relations between ratios form the simplest kind of interactions seen between the derived variables of plant growth analysis.

Where both parts of the ratio or fraction bear the same units, the quantity is a simple index of the importance of one component of the plant in relation to the whole. Ratios of this general type are: $R_W/W$, the root weight ratio (where $R_W$ is the total root dry weight of the plant); $S_W/W$, the stem weight ratio (where $S_W$ is the total stem dry weight of the plant); and $L_W/W$, the leaf weight ratio (where $L_W$ is the total leaf dry weight of the plant). The three are related by the expression.

$$R_W/W + S_W/W + L_W/W = 1.$$

Elias & Chadwick made a series of comparisons within this framework between 40 species and cultivars (Table 8.1, including reference).
An interrelation between more heterogeneous Type III quantities is seen during the subdivision of LAR,

$$L_A/W = L_A/L_W \times L_W/W$$

where $L_A/L_W$ is the specific leaf area, SLA, and $L_W/W$ is the leaf weight ratio, LWR. Looking simultaneously at all three, Jarvis & Jarvis established that the much less leafy nature of Scots pine, in comparison with sunflower, was due almost entirely to the relatively greater density of the pine needles and hardly at all to variation in LWR (the productive investment of the plant) which, in fact, showed a small difference in favour of pine (Table 8.2, including reference).

Equations used in an activity known as 'yield component analysis' also employ interrelations between Type III quantities. A typical equation relates the yield (as fruit dry weight per unit ground area) to its components as follows:

$$\text{Yield} = \text{Plant } n/\text{Ground area} \times \text{Fruit } n/\text{Plant } n \times \text{Fruit d.wt.}/\text{Fruit } n$$

where each $n$ is a number.

**Table 8.1 Leaf, stem and root weight ratios in ryegrass and clover** 95% confidence intervals are given in parenthesis. (From: C. O. Elias & M. J. Chadwick (1979). *J. Ecol.* 16, 537.)

| Species | Cultivar | LWR | SWR | RWR |
|---|---|---|---|---|
| *Lolium perenne* | S23 | 0.505 (0.023) | 0.245 (0.011) | 0.250 (0.028) |
| | S24 | 0.500 (0.017) | 0.238 (0.011) | 0.261 (0.019) |
| | Melle | 0.485 (0.022) | 0.273 (0.017) | 0.242 (0.014) |
| | Pelo | 0.512 (0.010) | 0.264 (0.019) | 0.224 (0.017) |
| | Standion | 0.512 (0.017) | 0.253 (0.018) | 0.235 (0.019) |
| | Endura | 0.497 (0.024) | 0.289 (0.020) | 0.214 (0.011) |
| *L. multiflorum* | Tiara | 0.547 (0.014) | 0.247 (0.024) | 0.206 (0.019) |
| *L perenne* × *L. multiflorum* | Sabrina | 0.534 (0.029) | 0.227 (0.024) | 0.239 (0.024) |
| *Trifolium repens* | S100 | 0.452 (0.036) | 0.316 (0.030) | 0.232 (0.022) |
| | Hiua | 0.458 (0.011) | 0.315 (0.022) | 0.266 (0.022) |
| | S184 | 0.483 (0.018) | 0.315 (0.027) | 0.202 (0.023) |
| | Blanca | 0.471 (0.019) | 0.303 (0.030) | 0.225 (0.017) |
| | Kersey | 0.444 (0.032) | 0.328 (0.031) | 0.228 (0.016) |
| | Pajbjerg | 0.457 (0.039) | 0.301 (0.035) | 0.242 (0.020) |
| *Trifolium pratense* | Tetri | 0.467 (0.016) | 0.274 (0.020) | 0.259 (0.012) |
| *Trifolium hybridum* | 'British' | 0.494 (0.013) | 0.277 (0.015) | 0.229 (0.018) |

**Table 8.2 Contributions of SLA and LWR to LAR in pine and sunflower** Instantaneous values. (Recalculated from P. G. Jarvis & M. S. Jarvis (1964). *Physiologia Pl.* 17, 654.)

| Species and harvest | LAR ($m^2\ g^{-1}$) | SLA ($m^2\ g^{-1}$) | LWR |
|---|---|---|---|
| *Pinus sylvestris*, Scots pine (at 2 years) | 0.0054 | 0.0084 | 0.643 |
| *Helianthus annuus*, sunflower (at *c.* 10 cm height) | 0.0234 | 0.0432 | 0.542 |

Relations between rates are very important in plant growth analysis because it is often useful to subdivide an index of overall performance, such as RGR, into indices which represent the individual performances of components of the system.

For example, the whole plant's relative growth rate in weight, $R_W$, can be subdivided into an expression which includes the relative growth rates of the individual organs of the plant. Hence:

$$\mathbf{R}_W = (1/W)(\mathrm{d}W/\mathrm{d}t) = (1/w_1)(\mathrm{d}w_1/\mathrm{d}t)w_1/W + (1/w_2)(\mathrm{d}w_2/\mathrm{d}t)w_2/W \cdots + (1/w_n)(\mathrm{d}w_n/\mathrm{d}t)w_n/W$$

where $w_1, w_2, \ldots, w_n$ are the dry weights of the individual parts of the plant, such as roots, stems and the successive leaves, in sum equalling $W$. Each of the terms on the right-hand side consists of the product of, first, the appropriate part's relative growth rate and, second, the ratio of that part to the whole, that is,

$$\mathbf{R}_W = \mathbf{R}_1 w_1/W + \mathbf{R}_2 w_2/W \cdots + \mathbf{R}_n w_n/W$$

where $\mathbf{R}_1, \mathbf{R}_2, \ldots, \mathbf{R}_n$ are the relative growth rates of $w_1, w_2, \ldots, w_n$. In practice, it is often more convenient to evaluate each of the right-hand terms as single quantities like $(1/W)(\mathrm{d}w_1/\mathrm{d}t)$, the rate of production of the component $w_1$ per unit of whole plant dry weight. This is a component production rate, with the symbol **J** (p. 70). Component production rates sum up to the whole plant relative growth rate, that is,

$$\mathbf{R}_W = \mathbf{J}_1 + \mathbf{J}_2 \cdots + \mathbf{J}_n$$

where $1 \ldots n$ represent the separate components, as before.

Data showing a complete set of component production rates obtained during the growth of nodal clusters of leaves of *Ambrosia trifida* (giant ragweed) obtained from a glasshouse experiment are given in Fig. 8.1.

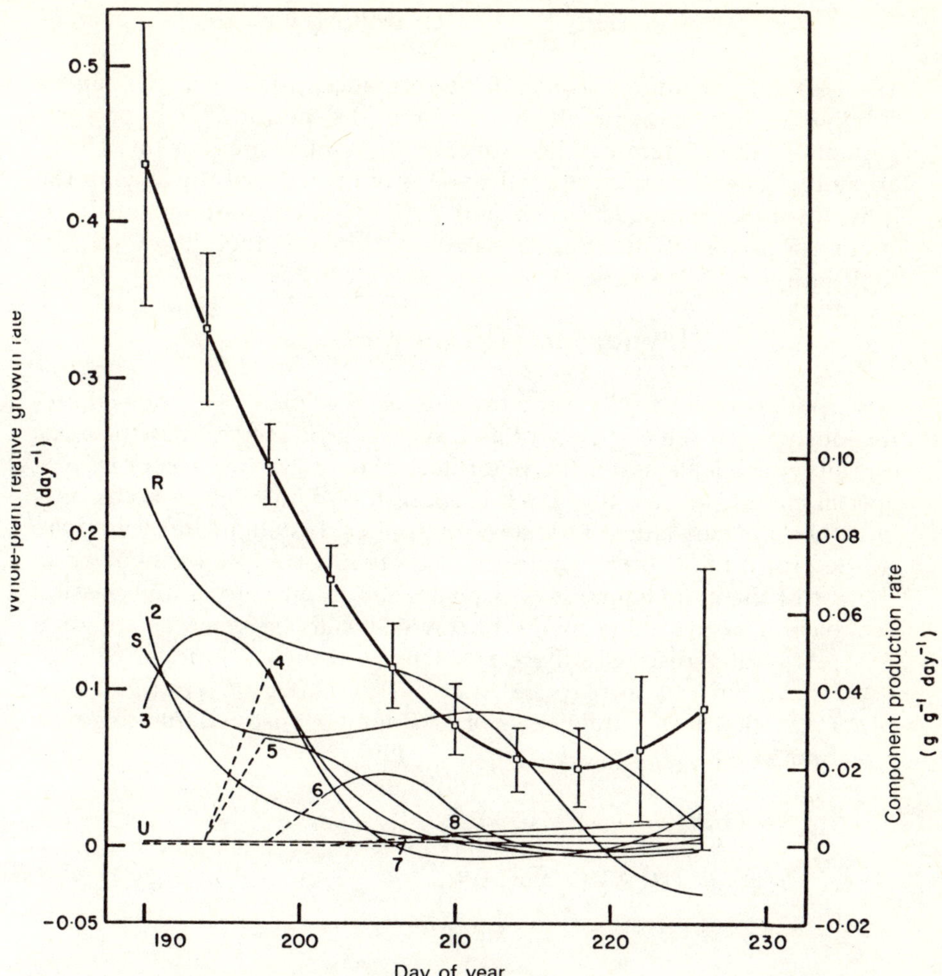


**Figure 8.1 Contribution of CPRs to RGR** The curves for component production rate in the different organs of giant ragweed: roots (R), stem material (S) and successive leaves (at nodes 1 to 8) (from Fig. 5.9). These rates sum up to a curve which describes the relative growth rate in dry weight of the whole plant (heavy line, with 95% limits). (From: R. Hunt & F. A. Bazzaz (1980). *New Phytol*. 84, 113.)

The idea that an index of overall performance, such as RGR, can be subdivided into indices which represent the performances of components of the system can be extended to involve subdivisions which are not wholly composed of further rates of growth. At the level of the individual plant, in fact, ULR and LAR evolved simultaneously as subdivisions of RGR. So, it is by definition that $\mathbf{R} = \mathbf{E} \times \mathbf{F}$ (or RGR = ULR × LAR) since

$$(1/W)(\mathrm{d}W/\mathrm{d}t) = (1/L_{\mathrm{A}})(\mathrm{d}W/\mathrm{d}t) \times L_{\mathrm{A}}/W.$$

Simply expressed, the growth rate of the plant depends simultaneously upon the efficiency of its leaves as producers of new material and upon the leafiness of the plant itself (Fig. 8.2). But, except in very special circumstances, $\bar{\mathbf{R}} \neq \bar{\mathbf{E}} \times \bar{\mathbf{F}}$ because the main relation holds only crudely for mean values of the three quantities. Instantaneous values are needed for it to be exact. There are also other assumptions involved in the use of the main equation which can lead to difficulty in the classical approach if plants are growing quickly or if harvest intervals are long. EVANS (p. 268) discusses these problems and their solutions.

We have already seen (p. 86) that $L_{\mathrm{A}}/W = L_{\mathrm{A}}/L_{\mathrm{W}} \times L_{\mathrm{W}}/W$ (or LAR = SLA × LWR). These subdivisions of LAR may be inserted into the main equation to give:

$$(1/W)(\mathrm{d}W/\mathrm{d}t) = (1/L_{\mathrm{A}})(\mathrm{d}W/\mathrm{d}t) \times L_{\mathrm{A}}/L_{\mathrm{W}} \times L_{\mathrm{W}}/W$$

otherwise, RGR = ULR × SLA × LWR.

**Figure 8.2 Contribution of LAR and ULR to RGR** These curves, for Kreusler's maize (Table 1.1), are reproduced (a) from Fig. 3.1, (b) from Fig. 4.1, and (c) from Fig. 5.1. Instantaneous values have been derived from fitted spline curves (see p. 105). The curve bearing 95% limits is for maize grown in 1875. Three other years' data are also shown: (○) 1876, (□) 1877 and (△) 1878. Instantaneously, RGR = LAR × ULR. The relative constancy of ULR against time means that RGR is principally controlled by the declining trend in LAR. (From: R. Hunt & G. C. Evans (1980). *New Phytol*. 86, 155.)

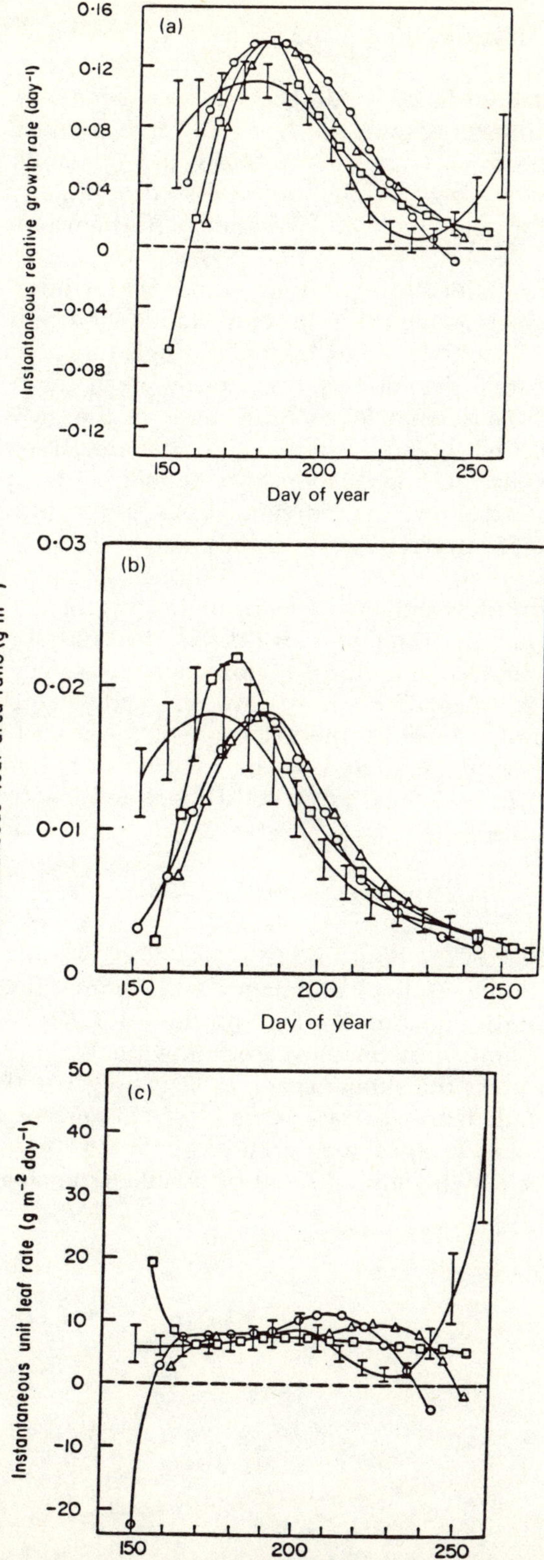

(a)
Instantaneous relative growth rate (day-1)
0·16
0·12
0·08
0·04
0
-0·04
-0·08
-0·12
150
200
250
Day of year
(b)
Instantaneous leaf area ratio (g m-2)
0·03
0·02
0·01
0
150
200
250
Day of year
(c)
Instantaneous unit leaf rate (g m-2 day-1)
50
40
30
20
10
0
-10
-20
150
200
250
Day of year

In the interrelations listed so far, the focus had been on plants grown as spaced individuals, despite the fact that large populations of similar individuals need to be raised to meet the demands of sequential destructive harvesting. Results are expressed on a per plant basis and the size of the population and its sum performance are of no direct concern.

However, in agriculture and in some field studies, where both populations (such as monospecific crop stands) and communities (such as mixed-species grassland) are treated as single functional units, we can also interpret their overall performances by means of interrelations. In one sense, these lie parallel to those used in the growth analysis of individual plants, because there is no theoretical reason why the concepts of RGR, ULR, LAR and so on, should not be applied on a per crop basis instead of on a per individual basis. But, in practice, the crop's parallel series of interrelations uses only one of these, unit leaf rate, in an unaltered form.

In the study of population and community growth, ULR provides exactly the same information as it does in the study of the growth of individuals, namely, an index of the functional efficiency of the productive parts of the plant. Straightforward reasoning suggests that overall yield is controlled both by the efficiency of the leaves of the crop as producers of dry material and by the leafiness of the crop itself. Since unit leaf rate, **E**, is defined as $(1/L_A)(\mathrm{d}W/\mathrm{d}t)$ and leaf area index, **L**, is $L_A/P$, we can write

$$(1/P)(\mathrm{d}W/\mathrm{d}t) = (1/L_A)(\mathrm{d}W/\mathrm{d}t) \times L_A/P.$$

This equation is $\mathbf{C} = \mathbf{E} \times \mathbf{L}$ (or CGR = ULR × LAI) and is the central relationship in the study of population and community growth, in the same way that the equation $\mathbf{R} = \mathbf{E} \times \mathbf{F}$ (or RGR = ULR × LAR) is central to the study of plants growing as spaced individuals.

We cannot press the similarity too closely, however. Though defined as a compounded growth rate, **C** is close in concept to an absolute growth rate; also, **L** and **F** may each be applied to the analysis of crop growth in their own right, instead of being the analogues suggested here.

Nevertheless, it is easy to see in broad terms that when comparing individuals with populations/communities, the overall performance of both systems (**C** and **R** respectively) may be broken down into two components: the productive efficiency of the leaves (**E** in both cases) and the leafiness (**L** and **F** respectively).

The calculation of the mean values **C**, **E**, and **L** all depend on various unrelated assumptions. Only rarely do these concur, so, in most cases, $\bar{\mathbf{C}} \neq \bar{\mathbf{E}} \times \bar{\mathbf{L}}$ for just the same type of reason that $\bar{\mathbf{R}} \neq \bar{\mathbf{E}} \times \bar{\mathbf{F}}$. In practice, this means that although the equation cannot properly be used to calculate any one quantity from mean values of the other two, the relationship is often close enough to be valuable in interpreting the overall growth of the system in terms of its component processes (e.g. Fig. 8.3).

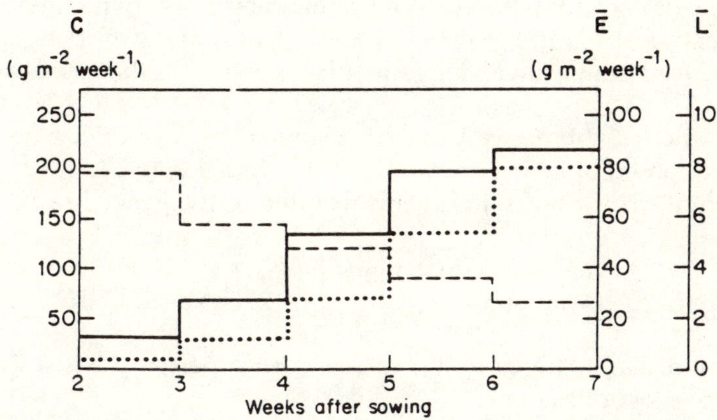


**Figure 8.3 Contribution of LAI and ULR to CGR** These progressions are of harvest-interval means of crop growth rate (C, solid line), unit leaf rate (E, dashed line), and leaf area index (L, dotted line) during the early growth of wheat. Instantaneously, C = E × L, so it is approximately the case that the decline with time in ULR is more than counterbalanced by the rise in LAI, leading to an increase in CGR. (From: V. Stoy (1965). *Physiologia Pl.* 18 (Suppl. IV), 1.)

Interrelations also exist which involve Type V quantities, the integral durations. For example if the LAD of a crop and its mean ULR are known then its final yield may be predicted. Less perversely, if this yield is already known (as it would be if ULR had been derived) then the yield may be broken down into its two components,

$$\underset{(MA^{-1})}{\text{Yield}} = \underset{(T)}{\text{LAD}} \times \underset{(MA^{-1}T^{-1})}{\text{ULR}}.$$

The approximation sign is used here because the concept of mean unit leaf rate over a whole season is inevitably very crude and because, in the concept of leaf area duration itself, equal areas beneath the LAI curve are treated as being equally useful opportunities for assimilation – an even more dubious assumption in view of the changes that take place, both ontogenetically within the crop and climatically in its environment, during the course of the crop's growth. EVANS (p. 224) discusses these problems further. An example involving four crops grown at Rothamsted appears in Table 8.3.

The other way to derive LAD is to estimate the area under the curve of $L_A$ (as opposed to LAI) versus time. This produces an LAD with units of area × time. This change in units is reflected in the general relationship

$$\underset{(M)}{\text{Yield}} \simeq \underset{(AT)}{\text{LAD (leaf area basis)}} \times \underset{(MA^{-1}T^{-1})}{\text{ULR}}.$$

This yield, like LAD, is now expressed on a per crop basis. The same inexactness applies here.

Another interrelation involves BMD, which has units of weight × time. The BMD is related to total yield, not by mean ULR, the productive efficiency of unit amounts of leaf area, but by the mean RGR itself, the productive efficiency of unit amounts of dry matter.

$$\underset{(M)}{\text{Yield}} \simeq \underset{(MT)}{\text{BMD}} \times \underset{([MM^{-1}]T^{-1})}{\text{RGR}}.$$

This relationship is also inexact but as a crude summary of the behaviour of the crop it can be useful.

**Table 8.3 Yield, leaf area duration and unit leaf rate in four Rothamsted crops** (From: D. J. Watson (1947). *Ann. Bot.* 11, 41.)

| Crop | Yield (tonne $ha^{-1}$) | LAD (weeks) | Mean ULR (tonne $ha^{-1}$ $week^{-1}$) |
|---|---|---|---|
| *Hordeum vulgare* (barley) | 7.3 | 17 | 0.43 |
| *Solanum tuberosum* (potato) | 7.7 | 21 | 0.36 |
| *Triticum* sp. (wheat) | 9.5 | 25 | 0.38 |
| *Beta vulgaris* (sugar beet) | 12.0 | 33 | 0.36 |

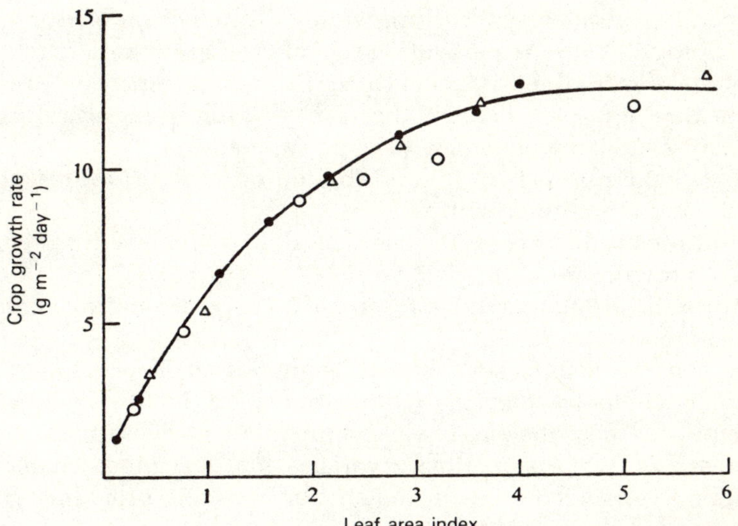


**Figure 8.4 Relation of CGR to LAI** Data for three equidistant spacings of soybean are shown: (•) 4.3, (○) 7.7, and (△) 12.9 plants $m^{-2}$. On the final plateau of the curve, increases in LAI are exactly counterbalanced by the decreases in ULR which arise from self-shading, so producing no net gain in CGR. (From: R. M. Shibles *et al.* (1975). In: L. T. Evans (ed.), *Crop Physiology: Some Case Histories*. Cambridge: Cambridge University Press.)

An elegant analysis by J. Warren Wilson (reference via Fig. 8.5) has not only linked time and environment (light flux density) together as determinants of plant size, but has also brought together the concepts involved in the growth analysis of individuals with those involved in populations and communities. Warren Wilson's 'universal' expression takes the form of a detailed subdivision of crop growth rate, notated as the product of the number of plants per unit area, $N$, and absolute growth rate per plant, $\mathrm{d}W/\mathrm{d}t$. Hence,

$$N\,\mathrm{d}W/\mathrm{d}t = \underset{(a)}{I_0} \times \underset{(b)}{N/I_0} \times \underset{(c)}{W} \times \underset{(d)}{L_W/W} \times \underset{(e)}{L_A/L_W} \times \underset{(f)}{J/L_A} \times \underset{(g)}{(1/J)(\mathrm{d}W/\mathrm{d}t)}$$

where $I_0$ is the incident light flux density and $J$ the light flux intercepted by the plant. The terms on the right-hand side of the equation may be grouped in various ways, each providing a subdivision of crop growth rate appropriate to one particular stage of crop growth.

First, taking (a), the product of (b) to (f), and (g) respectively provides the incident light flux density, the plant's 'intercepting efficiency' and the plant's 'utilizing efficiency' (a Type IV quantity).

Second, the products of (a) and (b) and of (c) to (g) represent plant density and absolute growth rate.

Third, the products of (a) to (c) and of (d) to (g) represent biomass and relative growth rate.

Fourth, the products of (a) to (e) and of (f) to (g) are leaf area index and unit leaf rate.

The scheme thus takes derived quantities of Types I to IV and combines them by means of a mathematical identity which owes something to the principles of yield component analysis (p. 86). Smooth progressions of the six primary variates may be fitted, enabling the equation to be solved on an instantaneous basis and providing progressions in all of the derived quantities.

Figure 8.5 provides an illustration of the first of these interrelations in action.

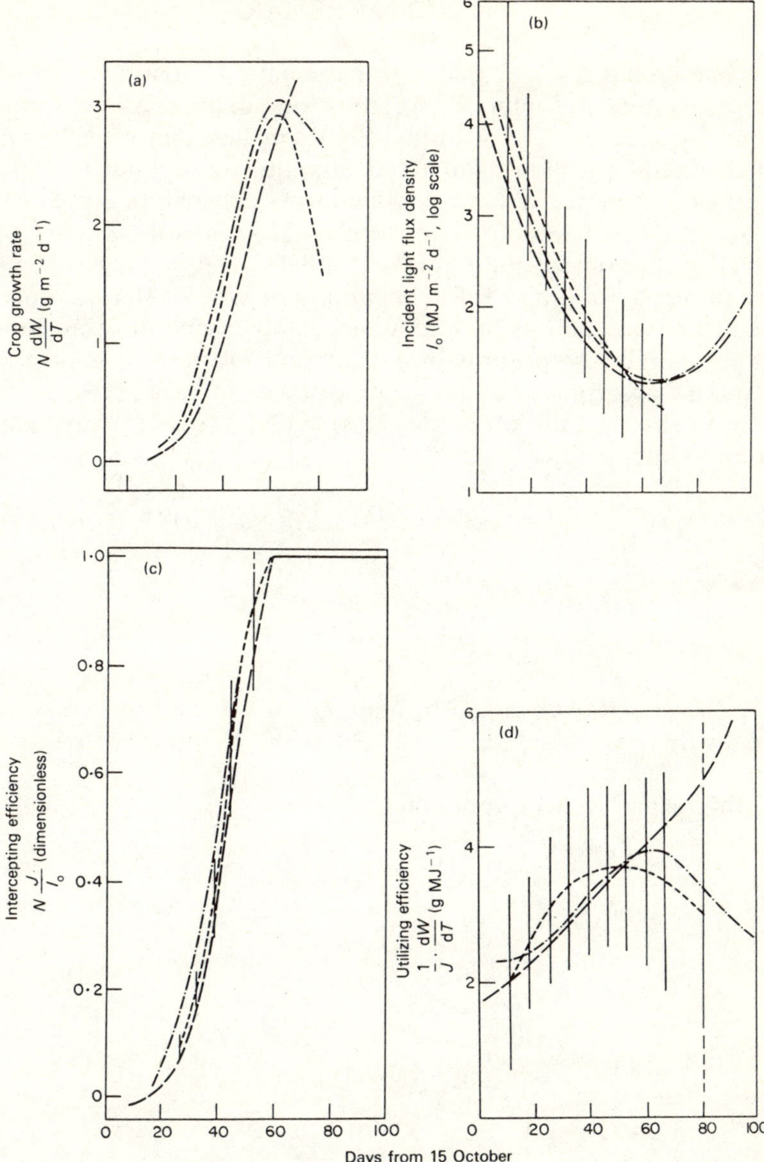


**Figure 8.5 Growth, light interception and light utilization** These four sets of curves are for winter lettuce planted under glass in mid-October, grown under $CO_2$ enrichment, and harvested in January. Three years' data are shown: 1978–79 (dotted lines, with 95% limits), 1979–80 (chain-dotted lines), 1980–81 (dashed lines). See also Fig. 4.4. The terms plotted are (a) crop growth rate, (b) incident light flux density, (c) intercepting efficiency, and (d) utilizing efficiency. The instantaneous relation illustrated is (a) = (b) × (c) × (d). (From: Hunt *et al.* (1984). *Ann. Bot.* 54, 743.)

## CONCLUDING REMARKS

This chapter has shown that the mathematical identities (pp. 84–85) which constitute the interrelating equations of plant growth analysis may take a great variety of forms. Some of these are trivially simple, involving only one type of derived quantity throughout (p. 86), and some are much more elaborate, involving a variety of types (p. 96); some form complementary and parallel counterparts with one another (pp. 90 and 92), and some are quite unique. (p. 88)

The principles outlined at the beginning of this chapter are universal to plant growth analysis and allow enormous scope for exansion. For example, in the case of growth and nutrient relations, it is possible to link specific absorption rate on a root dry weight basis, $(1/R_W)(dM/dt)$ (p. 62), to specific utilization rate, $(1/M)dW/dt)$ (p. 64). They bear this relation to one another

$$(1/R_W)(dM/dt) \times (1/M)(dW/dt) = (1/M)(dM/dt) \times (1/W)(dW/dt) \times W/R_W$$

otherwise

$$\mathbf{A} \times \mathbf{U} = \mathbf{R}_M \times \mathbf{R}_W \times W/R_W$$

where $\mathbf{R}_M$ and $\mathbf{R}_W$ are RGRs in mineral content and total dry weight respectively (pp. 28 and 26) and $W/R_W$ is the reciprocal of root weight fraction (p. 86).

Further examples will undoubtedly emerge.

# CHAPTER NINE

# *Questions and answers*

## QUESTIONS AND ANSWERS

The Natural Environment Research Council's Institute of Terrestrial Ecology publishes a series of *Statistical Checklists* (see Reference section). These aim to highlight the more significant questions to be taken into account during practical work, and they cover a number of different fields. By asking a series of 'loaded' questions they can provide a framework for marshalling thoughts and ideas on a subject.

The checklist entitled *Plant Growth Analysis* appeared in 1982. Here are the questions which it asked, with hints as to where some of the answers may be found.

### *Stating the objectives*

1 Have you stated clearly and explicitly the objectives of your research and the reasons for doing it? (see SATTLER p. 13.)

2 Have you translated these objectives into precise questions that the research may be expected to answer? (see ditto.)

### *Relevance of plant growth analysis*

3 Is your research expected to involve plant growth or decay, i.e. irreversible change in size or shape (however measured), or change in number? (See CAUSTON & VENUS p. 2; EVANS p. 11; HUNT p. 5.)

4 Do the questions you are asking require more detailed answers on plant form and function than would be available from studies in the fields of production ecology or plant demography? (see EVANS p. 421; HUNT p. 9.)

5 Do the questions you are asking require more broadly based answers on plant form and function than would be available from studies in the field of environmental physiology? (See HUNT p. 186.)

6 In particular, do you require an integrated assessment of the performance of whole organs, individuals, populations or communities? (See EVANS p. 516.)

7 Do you require this assessment to be made across ecologically or agronomically meaningful periods of time? (See ditto.)

8 Would it be useful to you to express plant performance in terms of derivates which are independent of the size of the system under study? (See CAUSTON & VENUS p. 17; EVANS p. 193; HUNT p. 16.)

9 Are any of the standard types of derivate in plant growth analysis useful to you either *per se* or as synthetic or comparative tools? For two generalized plant variates, $Y$ and $Z$, and for time $t$, these are:

| | | |
|---|---|---|
| absolute growth rates: | $\mathrm{d}Y/\mathrm{d}t$ | Type I |
| relative growth rates: | $(1/Y)(\mathrm{d}Y/\mathrm{d}t)$ | Type II |
| simple ratios: | $Y/Z$ | Type III |
| compounded growth rates: | $(1/Z)(\mathrm{d}Y/\mathrm{d}t)$ | Type IV |
| integral durations: | $\int_{t_1}^{t_2} Y\,\mathrm{d}t$ | Type V |

(See p. 16)

10 Are interrelations between these derivates, perhaps in the form of economic analogies, likely to be useful? For example,

$$(1/Y)(\mathrm{d}Y/\mathrm{d}t) = Z/Y \times (1/Z)(\mathrm{d}Y/\mathrm{d}t)$$
$$(\textit{`efficiency'}) = (\textit{`manpower'}) \times (\textit{`manpower productivity'}).$$

*Choice of approach*

11 Are you interested only in the net or final results of growth or decay over a longish period of time? (See HUNT p. 177.)

12 Are you interested in the detailed time-course of growth or decay, but know enough about this already to need only the level of definition provided by infrequent sampling? (See ditto.)

13 If the answer to 11 or 12 is 'yes', can you settle for the classical approach to plant growth analysis, in which mean values of derivates are obtained for the harvest-intervals between large, relatively infrequent samples? (See ditto.)

14 Are you unable to accept the assumptions implicit in the use of the classical harvest interval formulae (in particular, the problems of assembling accurate interrelationships as in 10)? (See EVANS p. 206.)

15 Do you know little or nothing about the time-course of the growth or decay you wish to study? (See MEAD & CURNOW p. 215.)

16 Do you require the most detailed picture possible, perhaps because you expect that species- or treatment-comparisons will be finely balanced? (See CAUSTON & VENUS p. 10; HUNT p. 51.)

17 If the answer to 14, 15, or 16 is 'yes', can you settle for the functional approach to plant growth analysis in which frequent, small harvests supply data for statistical curve-fitting? (See HUNT p. 51.)

18 Are any of the following requirements also convincing arguments for the functional approach? Great condensation of primary data; all sampling occasions in use for all comparisons; minimal physical risk or effort per harvest; unequal replication between harvests; unequal

harvest intervals; interpolation; smoothing; integral statistical analysis. (See HUNT p. 53.)

### *Experimental design and sampling*

19 In general, can you select the right experimental material and use it in such a way as to get down on a paper the best possible data for deriving answers to your original questions? (See EVANS p. 53.)

20 Have you read *Statistical Checklist No.1,* EVANS (p. 81) or MEAD & CURNOW (p. 33) on experimental design?

21 Have you read *Statistical Checklist No.2,* EVANS (p. 96) or MEAD & CURNOW (p. 15) on sampling?

22 Have you read EVANS (pp. 39–185) on measurement?

23 Have you examined the question of destructive versus non-destructive sampling? (See EVANS p. 40; HUNT p. 52.)

### *Inspection of data*

24 As plant variability is commonly value-dependent, do you agree that the natural logarithmic transformation, usually necessary for mathematical reasons, also renders the data homoscedastic, with Gaussian distributions within each harvest subset? (see CAUSTON & VENUS pp. 66, 133; HUNT pp. 56, 75.)

25 (Rarely) is such a transformation harmful to, or within, any of your data sets? (See ditto.)

### *Computing*

26 Have you investigated the locally available computing or calculating facilities to see what possibilities are available?

27 Have you read *Statistical Checklist No.3* on modelling (questions 71–76)?

28 Have you decided on whether to write your own program, use a library facility or borrow a program from another worker in the field?

29 If proceeding with any of 28, can you be sure of obtaining all of the necessary values, derivates and limits from your chosen method? (See HUNT p. 54.)

30 Do your computing facilities, as a whole, provide the best possible combination of availability, suitability and turnaround, or must some trade-off of one against another be achieved?

*Classical analytical methods*

31 Have you obtained the appropriate classical formulae, perhaps from a source such as EVANS, HUNT or here (p. 16).
32 Do you fully understand the assumptions involved in the use of each of these formulae? (See EVANS pp. 224, 255; HUNT p. 23.)
33 Can you pair your samples satisfactorily across each harvest interval? (See CAUSTON & VENUS p. 27; EVANS p. 440; HUNT p. 54.)
34 Can you assemble your classical derivates into populations which permit statistical analysis? (See CAUSTON & VENUS p. 27.)
35 Is there evidence of random instability in your classical derivates or can you genuinely believe what you see? (See HUNT p. 177.)

*Functional analytical methods*

36 Do you know where on the continuum between mechanism and empiricism your modelling (in the form of curve-fitting) is likely to lie? (See HUNT p. 47; FRANCE & THORNLEY p. 12.)
37 If you have firm mechanistic beliefs or hypotheses about the processes underlying the growth or decay that you are studying, can you select or devise a mathematical function which provides a convenient analogue of those processes? (See CAUSTON & VENUS p. 12; HUNT p. 185.)
38 Can you fit such a function to your data? (See CAUSTON & VENUS pp. 65, 86; HUNT p. 61.)
39 Can you then proceed to the required derivates, with statistical limits if possible? (See HUNT p. 54.)
40 If you have no mechanistic beliefs or hypotheses about your system but merely require a suitable statistical approximating function, can you select or devise a mathematical function which provides a convenient representation of your system? (See CAUSTON & VENUS p. 12; HUNT p. 185.)
41 Can you fit such a function to your data? (See CAUSTON & VENUS pp. 65, 86; HUNT p. 61.)
42 Can you then proceed to the required fitted derivates, with statistical limits if possible? (*See* HUNT p. 54.)
43 Are any of the following functions likely to be of value to you? (An exponential function is a function fitted to logarithmically transformed data.)

| | |
|---|---|
| First-order polynomial exponential: | for log-linear ('exponential') growth or decay; |
| Second-order polynomial exponential: | for simple-curving progressions; |

| | |
|---|---|
| Third-order polynomial exponentials: | For S-shaped progressions, or those with linearly changing curvature; |
| High-order polynomial exponentials: | for complex curves, but beware for unstable derivates and limits; |
| Monomolecular: | asymptotic, but without an inflection; |
| Logistic (autocatalytic): | asymptotic, with an inflection mid-way; |
| Gompertz: | asymptotic, with an early inflection; |
| Richards: | asymptotic, with a variable inflection; |
| Segments: | chains of simple functions fitted to complex data, but with 'lumpy' derivates; |
| Running re-fit: | ditto, with smooth derivates, but costly in degrees of freedom; |
| Splines: | specially joined polynomials, with seamless derivates and advantageously treated degrees of freedom; |
| Time series analysis: | for vast data sets with recurrent internal trends. |

(See HUNT chapters 5–7.)

44 Have you closely examined the question of determinate growth and of asymptotic versus non-asymptotic functions? (See CAUSTON & VENUS p. 86; HUNT p. 121.)
45 If used, have you sought incidental biological relevance in the parameters of the simpler functions listed in 43? (See HUNT p. 185.)
46 Have you balanced to your own satisfaction the often conflicting requirements of biological expectation and statistical exactitude? (See HUNT pp. 72, 105, 115, 159.)
47 Are you aware of the dangers of over-fitting your data? (See HUNT pp. 112, 157, 183.)
48 Are you aware of the dangers of under-fitting your data? (See HUNT pp. 72, 83, 96, 104, 112, 157, 162.)
49 Have you used progressions of the derivates as incidental or alternative indications of goodness or suitability of fit? (See HUNT p. 157.)

50 For low-order polynomials, is the stepwise principle of any value? (See HUNT. p. 110.)
51 If it is, can you accept a variety of types of fit within your collection of data sets, or must they all be similarly treated for any reason? (See HUNT p. 114.)

*Post-analysis: assessment and presentation*

52 Have you applied 'Occam's razor' wherever possible (accepting the simplest of alternative explanations)? (See HUNT p. 115.)
53 Has your analysis created a 'Procrustean bed' (an abuse of certain data sets in pursuit of overall uniformity)? (See EVANS p. 4.)
54 Have you got everything you wanted from the analysis?
55 Must you return to another method, perhaps because of the unavailability of meaningful derivates or statistical limits?
56 If your derivates are time-based, can you gain additional information by plotting them not against time but against some other measure of progress, such as total dry weight? (See EVANS p. 449; HUNT p. 40.)
57 Can your incorporate such an alternative measure directly into any of the calculations? (See HUNT p. 40.)
58 If you have experimental treatments based not upon multi-state but upon continuous variables, can you incorporate either the primary or derived data into a response surface involving independent variables in addition to time? (See HUNT p. 168.)
59 If your original design admits both classical and functional methods of analysis, can additional information be gained by performing both?
60 Will you present each classical rate-derivative not as single points on a progression in time, but as a histogram with a class-interval equal to the harvest-interval? (See EVANS p. 212; HUNT pp. 15, 20.)
61 Will you need to present all of the statistical limits calculated for fitted derivates or can economy of presentation be achieved by giving only the 'control', or perhaps the widest, set of limits? (See HUNT p. 159.)
62 Can values of derivates from empirical work be used to set values or limits to state variables or rate equations in subsequent mechanistic work? (See HUNT p. 49; FRANCE & THORNLEY p. 75.)

*The final (and most important) question*

63 If you are in any doubt about the purpose of any of the questions in this checklist, should you not obtain some advice from a worker with experience of plant growth analysis before continuing? There is usually little that an expert advisor can do to help you once you have committed yourself to a faulty approach.

# *Tables of synopses*

**Synopsis of the main derived quantities**

| Quantity | Contraction | Symbol |
|---|---|---|
| Absolute growth rate | AGR | **G** |
| Allometric coefficient | — | **K** |
| Biomass duration | BMD | **Z** |
| Component production rate | CPR | **J** |
| Crop growth rate | CGR | **C** |
| Leaf area duration (leaf area basis) | LAD | **D** |
| Leaf area duration (leaf area index basis) | LAD | **D** |
| Leaf area index | LAI | **L** |
| Leaf area ratio | LAR | **F** |
| Leaf weight ratio | LWR | — |
| Relative growth rate | RGR | **R** |
| Specific absorption rate | SAR | **A** |
| Specific leaf area | SLA | — |
| Specific utilization rate | SUR | **U** |
| Sub-cellular efficiency | — | $\mathbf{G}_{i,i-1}$ |
| Unit leaf rate (net assimilation rate) | ULR (NAR) | **E** |
| Unit production rate | UPR | **Π** |
| Unit shoot rate | USR | **B** |

| Definition | 'Type' | Dimension | See pp. |
|---|---|---|---|
| $\mathrm{d}W/\mathrm{d}t$, $\mathrm{d}N/\mathrm{d}t$ | I | M $T^{-1}$, N $T^{-1}$ | 18–24, 96–97 |
| $(\mathrm{d}(\log w_1)/\mathrm{d}t)/(\mathrm{d}(\log w_2)/\mathrm{d}t)$ | III | dimensionless | 50–53 |
| $_{t1}\int^{t2} W\,\mathrm{d}t$ | V | M T | 76–77, 94 |
| $(1/W)(\mathrm{d}w_1/\mathrm{d}t)$ | IV | M $M^{-1}$ $T^{-1}$ | 70–71, 88–89 |
| $(1/P)(\mathrm{d}W/\mathrm{d}t)$ | IV | M $A^{-1}$ $T^{-1}$ | 60–61, 92–97 |
| $_{t1}\int^{t2} L_A\,\mathrm{d}t$ | V | A T | 74 |
| $_{t1}\int^{t2} L\,\mathrm{d}t$ | V | T | 74–75, 94–95 |
| $L_A/P$ | III | dimensionless | 42–43, 92–93 |
| $L_A/W$ | III | A $M^{-1}$ | 36–37, 90–91 |
| $L_W/W$ | III | dimensionless | 40–41, 86–87 |
| $(1/W)(\mathrm{d}W/\mathrm{d}t)$ | II | [M $M^{-1}$] $T^{-1}$ | 26–34, 88–91 |
| $(1/R_w)(\mathrm{d}M/\mathrm{d}t)$ | IV | M $M^{-1}$ $T^{-1}$ | 62–63 |
| $L_A/L_W$ | III | A $M^{-1}$ | 38–39, 90–91 |
| $(1/M)(\mathrm{d}W/\mathrm{d}t)$ | IV | M $M^{-1}$ $T^{-1}$ | 64–65 |
| $(1/w_{i-1})(\mathrm{d}w_i/\mathrm{d}t)$ | IV | M $M^{-1}$ $T^{-1}$ | 66–67 |
| $(1/L_A)(\mathrm{d}W/\mathrm{d}t)$ | IV | M $A^{-1}$ $T^{-1}$ | 56–59, 90–95 |
| $(1/W_p)(\mathrm{d}W/\mathrm{d}t)$ | IV | M $M^{-1}$ $T^{-1}$ | 68–69 |
| $(1/S_W)(\mathrm{d}W/\mathrm{d}t)$ | IV | M $M^{-1}$ $T^{-1}$ | 58–59 |

**Synopsis of primary quantities and associated terms**

| Symbol | Quantity | Dimension |
|---|---|---|
| $a$ | Generalized mathematical variable | as required |
| $b$ | Ditto; intercept coefficient in allometry | as required |
| $f$ | Generalized mathematical function | as required |
| $f'$ | Slope of $f$ | as required |
| $I_0$ | Incident light flux density | $H\ T^{-1}$ |
| $I_1$, $I_2$ | Total input and output energies | H |
| $J$ | Intercepted light flux density | $H\ T^{-1}$ |
| $k$ | Carrying capacity | N |
| $L$ | Generalized total size of leaf material | as required |
| $L_A$ | Total leaf area per plant or per crop | A |
| $L_W$ | Total leaf dry weight per plant | M |
| $M$ | Total mineral nutrient content per plant (comprising one or more elements) | M |
| $n$ | Generalized number | N |
| $N$ | Number of individuals; planting density | number: $A^{-1}$ |
| $P$ | Ground area per sample; plot area | A |
| $r$ | Intrinsic rate of increase | $T^{-1}$ |
| $R$ | Root size | as required |
| $R_W$ | Total root dry weight per plant | M |
| $S_W$ | Total shoot dry weight per plant (above-ground parts of the plant); total stem dry weight per plant (according to context) | |
| $t$ | Time | T |
| $W$ | Total dry weight per plant or per crop | M |
| $w_i$ | Dry weight of the *i*th plant/cell component | M |
| $W_M$ | Marketable dry weight per crop | M |
| $W_P$ | Total dry weight of perennating material | M |
| $W_{pr}$ | Weight of protein per crop | M |
| $x$, $y$, $z$ | Generalized mathematical variables | as required |
| $Y$, $Z$ | Generalized plant variates | as required |

# *References*


ALLEN, S. E. (ed.) 1974. *Chemical analysis of ecological materials*. Oxford: Blackwell Scientific Publications.
(For analyses of mineral nutrient contents and inorganic constituents.)

CAUSTON, D. R. 1983. *A biologist's basic mathematics*. London: Edward Arnold.
(For the mathematics of plant growth.)

CAUSTON, D. R. & J. C. VENUS 1981. *The biometry of plant growth*. London: Edward Arnold.
(For the functional approach, regression theory and the growth of components.)

COOMBS, J. & D. O. HALL 1982. *Techniques in bioproductivity and photosynthesis*. Oxford: Pergamon.
(For production ecology and biotechnology.)

EVANS, G. C. 1972. *The quantitative analysis of plant growth*. Oxford: Blackwell Scientific Publications.
(For history, experimental technique and the classical approach.)

FRANCE, J. & J. H. M. THORNLEY 1984. *Mathematical models in agriculture: a quantitative approach to problems in agriculture and related sciences*. London: Butterworth.
(For the philosophy, techniques and applications of dynamic, deterministic modelling.)

HUNT, R. 1982. *Plant growth curves: the functional approach to plant growth analysis*. London: Edward Arnold.
(For the philosophy, techniques and applications of the functional approach.)

JEFFERS, J. N. R. 1979. *Sampling: Statistical Checklist No. 2*. Institute of Terrestrial Ecology: Grange-over-Sands.
(For the principles and practice of statistical sampling.)

JEFFERS, J. N. R. 1980. *Modelling: Statistical Checklist No. 3*. Institute of Terrestrial Ecology: Grange over-Sands.
(For the rationale and design of formal modelling.)

JEFFERS, J. N. R. 1984. *Design of experiments: Statistical Checklist No. 1*. Institute of Terrestrial Ecology: Grange-over-Sands.
(For the principles and practice of experimental design.)

MEAD, R. & R. N. CURNOW 1983. *Statistical methods in agriculture and experimental biology*. London: Chapman and Hall.
(For experimental design, significance testing and regression theory.)

SATTLER, R. 1986. *Biophilosophy: analytic and holistic perspectives*. Berlin: Springer-Verlag.
(For the theoretical and philosophical foundations of plant science.)
SILVERTOWN, J. W. 1982. *Introduction to plant population ecology*. London: Longman.
(For population dynamics and plant demography.)